Number Theory - MOD

Math All Star Practice by Subject Series

Math for Gifted Students

Copyright © 2015 by MathAllStar. All rights reserved.

No part of this book may be reproduced, distributed or transmitted in any form or by any means, including photocopying, scanning, or other electronic or mechanical methods, without written permission of the author.

To promote education and knowledge sharing, reuse of some contents of this book may be permitted, courtesy of the author, provided that: (1) the use is reasonable; (2) the source is properly quoted; (3) the user bears all responsibility, damage and consequence of such use. The author hereby explicitly disclaims any responsibility and liability; (4) the author is notified in advance; and (5) the author encourages, but does not enforce, the user to adopt similar policies towards any derived work based on such use.

Please visit the website `https://www.mathallstar.org` for more information or email `contact@mathallstar.org` for suggestions, comments, questions and all copyright related issues.

use your mobile device to scan this QR code for more resources including books, practice problems, online courses, and blog.

This book was produced using the LaTeX system.

How to get the most out of practice

The most important tip is to learn before practice. This approach is what students do in schools. It also applies to competition math. Systematic learning will help students develop the following skills, and get the most out of practice afterwards.

1. Be able to recognize which subject a problem belongs to.
2. Know relevant solving techniques for such type of problems.
3. Be able to choose the most appropriate solution for this particular problem.

There are many tutorial materials available on Math All Star website, including books, videos, articles, and so on. The address is `https://www.mathallstar.org`.

Meanwhile, it is important to read and understand reference solution instead of just checking the answer. One objective of practice is for students to check whether they understand and master all the necessary solving techniques or not. However, merely obtaining the correct answer does not necessarily mean the most suitable technique is used. Therefore, it is beneficial to understand the solution in addition to obtaining the correct answer.

Contents

I Review   1

1 MOD Basic   3
  1.1 MOD Basic 3
      1.1.1 Definition 3
      1.1.2 Residue Class and Residue System 4
  1.2 MOD Operations 4
      1.2.1 Basic Properties 4
      1.2.2 Modular Multiplicative Inverse 5
  1.3 The Negative One Method 6
  1.4 Sum of Digits (The MOD 9 Technique) 7
  1.5 Finding Ending Digit(s) 8
      1.5.1 Finding the Units Digit 8
      1.5.2 Finding the Last k Digits 9
      1.5.3 A Quick Way to Find the Tens Digit of m^n . . 10
      1.5.4 Useful Facts 13
  1.6 Square Numbers 13
      1.6.1 MOD Properties 13
      1.6.2 Sum of Squares 14

2 Important Theorems   17
  2.1 Fermat Little Theorem 17
  2.2 Euler's Totient Function 18
  2.3 Euler's Theorem 19
  2.4 Multiplicative Order 21
  2.5 Wilson's Theorem 22

3 MOD Equation   23
  3.1 MOD Equation Basics 23

3.2 Solving $ax \equiv b \pmod{m}$ 25
3.3 Chinese Remainder Theorem (CRT) 26

II Practice   29

III Solution   69

Part I

Review

Chapter 1

MOD Basic

1.1 MOD Basic

1.1.1 Definition

Let a, b, and m be three integers where $m > 0$. If the difference of a and b is a multiple of m, we say a *is congruent to b modulo m* and write this relationship as

$$a \equiv b \pmod{m}$$

For example,

$$30 \equiv 16 \equiv 2 \equiv -5 \pmod{7}$$

It is important to note that a and b can be negative. In fact, using negative numbers is a frequently employed technique to simplify computation.

1.1.2 Residue Class and Residue System

A *residue class* is a set of all integers which are congruent to each other modulo m. For modulo m, there are exactly m different residue classes which are congruent to 0, 1, $\cdots$, and $(m-1)$, respectively.

A *complete residue system* is a set of integers which contains exactly one element from each of all the residue classes. For modulo m, it is clear that a complete residue system contains exactly m elements. There are an infinite number of complete residue systems for modulo m. The following is called the least non-negative residue system:
$$\{0, 1, \cdots, m-1\}$$

1.2 MOD Operations

1.2.1 Basic Properties

MOD operations follow the same basic rules as regular algorithmic except the division operation.

Theorem 1.2.1 MOD Addition and Subtraction

Let $a \equiv b \pmod{m}$ and $c \equiv d \pmod{m}$, then
$$a \pm c \equiv b \pm d \pmod{m}$$

Theorem 1.2.2 MOD Multiplication

Let $a \equiv b \pmod{m}$ and $c \equiv d \pmod{m}$, then
$$a \times c \equiv b \times d \pmod{m}$$

> **Theorem 1.2.3 MOD Multiplication with Constant**
>
> Let $a \equiv b \pmod{m}$ and k be an integer, then
> $$k \times a \equiv k \times b \pmod{m}$$

> **Theorem 1.2.4 MOD Exponentiation**
>
> Let $a \equiv b \pmod{m}$ and k be a positive integer, then
> $$a^k \equiv b^k \pmod{m}$$

It is important to note the division is an exception. In order to make division work, the relationship between the divisor k and the modulo m must be taken into consideration.

> **Theorem 1.2.5 MOD Division**
>
> Let $a \equiv b \pmod{m}$ and k be an integer, then
> $$\frac{a}{k} \equiv \frac{b}{k} \pmod{\frac{m}{\gcd(m,k)}}$$
> where $\gcd(m, k)$ is the greatest common divisor of m and k.

For example, given $30 \equiv 6 \pmod{8}$, simply dividing both sides by 2 will invalidate the congruence relationship: $15 \not\equiv 3 \pmod{8}$. Instead, it is necessary to divide the modulo by 2 which is the greatest common divisor of 8 and 2: $15 \equiv 3 \pmod{4}$.

1.2.2 Modular Multiplicative Inverse

Given an integer a, its modular multiplicative inverse modulo m is also an integer which is written as a^{-1} or $\frac{1}{a}$ and satisfies the relation $aa^{-1} \equiv 1 \pmod{m}$.

CHAPTER 1. MOD BASIC

> **Theorem 1.2.6   Existence of Modular Multiplicative Inverse**
>
> The modulo multiplicative inverse of a modulo m exists if and only if a and m are co-prime.

If a^{-1} exists, it must be one of $1, 2, \cdots, (m-1)$. It can be shown that the modular multiplicative inverse is unique within this range. Therefore, for a small integer m, the easiest way to find a^{-1} is just to go through all the values within this range. For example, $\frac{1}{3} \equiv 5 \pmod 7$ because $3 \times 5 \equiv 1 \pmod 7$. No other integer within $[1, 6]$ satisfies this condition.

Multiplicative inverse can also be computed using the Euclid algorithm or Euler's theorem. Euclid algorithm is a standard way to solve $ax + by = 1$ which is discussed in the book *Indeterminate Equation* in the Math All Star series. Euler's theorem will be covered later.

1.3   The Negative One Method

The negative one method is one of the most used elementary techniques to evaluate a MOD expression.

Example 1.3.1

Find all positive integers n such that $2^n + 1$ is divisible by 3.

Solution

The answer is the set of all odd integers because

$$2^n + 1 \equiv (-1)^n + 1 \equiv 0 \pmod 3$$

holds if and only if n is odd.

CHAPTER 1. MOD BASIC

Done.

Sometimes, a problem may require a few intermediate steps to obtain (-1). For example,

$$3^{2018} \equiv (3^2)^{1009} \equiv 9^{1009} \equiv (-1)^{1009} \equiv -1 \equiv 9 \pmod{10}$$

Similarly, obtaining a positive one as the base is also useful. In addition to outrightly get a positive one such as $10 \equiv 1 \pmod 9$, Fermat's little theorem and Euler's theorem can be effective in some complicated cases. Both of them will be discussed later.

1.4 Sum of Digits (The MOD 9 Technique)

When a problem involves the sum of digits, this technique should be considered.

> **Theorem 1.4.1 Sum of Digits**
>
> Let n be a positive integer and $S(n)$ be the sum of its digits, then
> $$n \equiv S(n) \pmod 9$$

The 18th problem in 2017 AMC12A offers a typical example.

Example 1.4.1

Let $S(n)$ equal the sum of the digits of positive integer n. For example, $S(1507) = 13$. For a particular positive integer n, $S(n) = 1274$. Which of the following could be the value of $S(n+1)$?

(A) 1   (B) 3   (C) 12   (D) 1329   (E) 1265

CHAPTER 1. MOD BASIC

Solution

By the MOD 9 technique, we have $n \equiv S(n) \equiv 1274 \equiv 5 \pmod{9}$. Therefore $n + 1 \equiv 6 \pmod{9}$. Among the given choices, only 1329 satisfies this constraint.

Done.

Another type of problem which can be solved with this technique is to find a missing digit while all the others present. An example is shown below:

Example 1.4.2

The number 2^{29} is a nine-digit number whose digits are all distinct. Which digit of 0 to 9 does not appear?

A solution to this example is provided in the practice.

1.5 Finding Ending Digit(s)

1.5.1 Finding the Units Digit

Finding the units digit of m^n is usually easy. The result only depends on the units digit of m. An elementary method to find the last digit is just to observe the repeating pattern. For example, the last digits of 7^n where $n = 1, 2, \cdots$ are

$$7, 9, 3, 1, 7, 9, 3, 1, \cdots$$

Therefore, evaluating $n \pmod{4}$ can help determine the last digit of 7^n.

CHAPTER 1. MOD BASIC

1.5.2 Finding the Last k Digits

Finding the last k digits of m^n is equivalent to evaluating m^n (mod 10^k). In addition to the usual calculation techniques, binomial expansion is a powerful method to find the last k digits when m ends with one or nine.

Example 1.5.1

What are the last 3 digits of 2019^{2019}?

Solution

Firstly, $2019^{2019} \equiv 19^{2019}$ (mod 1000). Next, let's rewrite 19 as $(20-1)$ and expand the expression:

$$(20-1)^{2019} = \cdots + C_{2019}^3 \times 20^3 - C_{2019}^2 \times 20^2 + C_{2019}^1 \times 20 - 1$$

It is clear that all the terms except the last three will be multiples of 1000 whose last 3 digits are always 000. Therefore, it is sufficient to calculate the last three digits of the sum of the last three terms.

$$-C_{2019}^2 \times 20^2 + C_{2019}^1 \times 20 - 1$$
$$= -\frac{2019 \times 2018}{2} \times 400 + 2019 \times 20 - 1$$
$$\equiv -2019 \times 2018 \times 200 + 2019 \times 20 - 1$$
$$\equiv -19 \times 18 \times 200 + 19 \times 20 - 1$$
$$\equiv -21 \equiv 979 \pmod{1000}$$

Hence, the answer is $\boxed{979}$.

Done.

CHAPTER 1. MOD BASIC

1.5.3 A Quick Way to Find the Tens Digit of m^n

In addition to the general method of finding the last k digits of m^n, a quick method exists to find the last two digits. Because the units digit is alway easy to determine, the key is to decide the tens digit.

<u>m ends with 0 or 5</u>

When m ends with 0, then the tens digit of m^n is always 0 when $n \geq 2$. When m ends with 5 and greater than 10, it can be shown that m^n can only end with 25 or 75. These two cases are trivial. No special trick is need.

<u>m ends with 1</u>

This case can be tackled with the following technique.

> **Theorem 1.5.1** Finding tens digit of m^n when m ends with 1
>
> Take the tens digit of m and multiply it with the units digit of n, then their product's units digit is the tens digit of m^n.

Let's consider the following example.

Example 1.5.2

What is the tens digit of 321^{123}?

Solution

The tens digit of 321 is 2. The units digit of 123 is 3. Thus, the tens digit of 321^{123} is $2 \times 3 = \boxed{6}$.

Done.

In another word, the last two digits of 321^{123} is 61, or $321^{123} \equiv 61$ (mod 100).

Theorem 1.5.1 is the steppingstone of finding the tens digit when m ends with other digits. This theorem can be proved using binomial expansion which is described in the previous section.

<u>m ends with 3, 7, or 9</u>

These three cases can be transformed to the previous case where m ends with 1 by noting that 3^4, 7^4, and 9^2 all end with 1. Here is an example.

Example 1.5.3

Find the last two digits of 123^{321}.

Solution

Firstly, we note that
$$123^{321} \equiv 23^{321} = \left(23^4\right)^{80} \times 23^1 \pmod{100}$$

Because 23^4 ends with 1, the units digit of $\left(23^4\right)^{80}$ must be 1. By *Theorem 1.5.1*, the tens digit of $\left(23^4\right)^{80}$ must be 0 because the units digit of the exponent is 0. In another word, $\left(23^4\right)^{80} \equiv 01$ (mod 100). Setting this to the above relation leads to
$$123^{321} \equiv 01 \times 23 = \boxed{23} \pmod{100}$$

Done.

It is important to note that 3, 7, and 9 all relate to certain powers of 3:
$$3^1 = 3, 3^2 = 9, 3^3 = 27$$

Therefore, the technique discussed in this section can help to determine the tens digit of a number in the form of 3^k.

CHAPTER 1. MOD BASIC

Other Cases

Tackling the remaining cases requires a bit of creative thinking. The essential technique is to first factorize m and then process each part separately. A prime factor can only be 2 or ends with 1, 3, 5, 7, or 9. We have already discussed those cases when m ends with an odd digit. Therefore, the only scenario left to tackle is 2^k where k is a positive integer.

The technique to find the tens digit of 2^k depends on the following facts:

- $2^{10} = 1024$, ends with 24

- $24^2 = 576$, ends with 76

- Any power of 76 always ends with 76 itself

This means that if k is sufficiently large, 2^k can be rewritten as a product of a smaller power of 2 and 76^p, modulo 100.

Let's review an example.

Example 1.5.4

Determine the last two digits of 312^{123}.

Solution

First, let's factorize the expression to:
$$312^{123} \equiv 12^{123} = (2^2 \times 3)^{123} = 2^{246} \times 3^{123} \pmod{100}$$
and then tackle the two terms separately:
$$2^{246} = (2^{10})^{24} \times 2^6 \equiv 24^{24} \times 64 \equiv (24^2)^{12} \times 64 \equiv 76 \times 64 \pmod{100}$$
$$3^{123} = (3^4)^{30} \times 3^3 = 81^{30} \times 27 \equiv 01 \times 27 = 27 \pmod{100}$$

Therefore, the answer we are looking is
$$76 \times 64 \times 27 \equiv \boxed{28} \pmod{100}$$

Done.

1.5.4 Useful Facts

- 25 and 76 are the only two-digit integers whose k^{th} powers always end with themselves where k is a positive integer.

- 376 and 625 are the only three-digit integers whose k^{th} powers always end with themselves where k is a positive integer.

- 7^{4k} always ends with 01 where k is a positive integer.

1.6 Square Numbers

1.6.1 MOD Properties

Many square number related problems can be solved using the MOD method. One of the most frequently used properties is shown below.

Theorem 1.6.1 Square Number MOD Property

Let n^2 be a square number, then $n^2 \equiv 0, 1 \pmod{4}$.

Proof

If n is odd, then $n \equiv \pm 1 \pmod{4} \implies n^2 \equiv 1 \pmod{4}$.

If n is even, then $n \equiv 0, 2 \pmod{4} \implies n^2 \equiv 0 \pmod{4}$

Hence, we always have $n^2 \equiv 0, 1 \pmod{4}$.

QED

CHAPTER 1. MOD BASIC

Additional properties can be derived using a similar approach. For example:

$$n^2 \equiv 0, \pm 1 \pmod{5}$$
$$n^2 \equiv 0, 1, 4 \pmod{8}$$
$$n^2 \equiv 0, 1, 4, 7 \pmod{9}$$

A useful extension to *Theorem 1.6.1* is illustrated in the following example.

Example 1.6.1

Find all the integer pairs (x, y) such that $x^2 + y^2 = 2019$.

Solution

No such pair exists. This is because

$$x^2, y^2 \equiv 0, 1 \pmod{4} \implies x^2 + y^2 \not\equiv 3 \pmod{4}$$

However, $2019 \equiv 3 \pmod{4}$.

Done.

1.6.2 Sum of Squares

Problems related to sum of squares are often associated with MOD too.

Theorem 1.6.2   Sum of Squares

A prime $p > 2$ is a sum of two squares if and only if $p \equiv 1 \pmod{4}$.

This theorem can be further generalized as follows to include composites.

> **Theorem 1.6.3   Two Squares Theorem**
>
> A positive integer n is a sum of two squares if and only if each prime factor p of n such that $p \equiv 3 \pmod 4$ occurs to an even power in the prime factorization of n.

Proofs of these two theorems are given in the practice.

CHAPTER 1. MOD BASIC

Chapter 2

Important Theorems

2.1 Fermat Little Theorem

This theorem is stated as follows:

> **Theorem 2.1.1 Fermat Little Theorem**
> Let p be a prime and a be an integer, then
> $$a^p \equiv a \pmod{p} \qquad (2.1)$$

While all of the following conclusions can be proved with elementary methods, they can be derived directly by the Fermat's little theorem.

Example 2.1.1

For any positive integer x, we must have

$$x^2 \equiv x \pmod{2}$$
$$x^3 \equiv x \pmod{3}$$
$$x^5 \equiv x \pmod{5}$$
$$x^7 \equiv x \pmod{7}$$

When a and p are co-prime, we can divide both sides of this congruence by a thus to obtain a 1. This can be used as a *positive one technique* which is similar to the negative one technique discussed early in *Section 1.3*.

Because p is a prime, a is co-prime to p is equivalent to $p \nmid a$. In this case, Fermat's little theorem can be expressed as follows.

Theorem 2.1.2 Fermat Little Theorem Alternative

Let p be a prime and a be an integer not divisible by p, then

$$a^{p-1} \equiv 1 \pmod{p} \qquad (2.2)$$

2.2 Euler's Totient Function

Given a positive integer n, Euler's totient function, written as $\varphi(n)$, returns the number of positive integers less than n which are co-prime to n. By employing the inclusion-exclusion principle[1], we can derive the following formula.

---

[1]This principle is discussed in the book *Counting* in the Math All Star series.

Theorem 2.2.1 Euler's Totient Function

Let positive integer n's prime factorization be $n = p_1^{\alpha_1} p_2^{\alpha_2} \cdots p_k^{\alpha_k}$, then

$$\varphi(n) = n \left(1 - \frac{1}{p_1}\right)\left(1 - \frac{1}{p_2}\right) \cdots \left(1 - \frac{1}{p_k}\right)$$

or

$$\varphi(n) = p_1^{\alpha_1 - 1} p_2^{\alpha_2 - 1} \cdots p_k^{\alpha_k - 1} (p_1 - 1)(p_2 - 1) \cdots (p_k - 1)$$

For example, because $100 = 2^2 \times 5^2$, we find

$$\varphi(100) = 100 \times \left(1 - \frac{1}{2}\right) \times \left(1 - \frac{1}{5}\right) = 40$$

This means that there are 40 positive integers less than 100 are co-prime to 100.

Euler's totient function has the following basic properties:

- Let m and n be two co-prime positive integers. Then $\varphi(mn) = \varphi(m)\varphi(n)$.

- Let p be a prime, then $\varphi(p) = p - 1$.

- Let p be a prime and k be a positive integer, then $\varphi(p^k) = p^{k-1}(p - 1)$.

2.3 Euler's Theorem

Fermat Little Theorem can be generalized into Euler's Theorem.

CHAPTER 2. IMPORTANT THEOREMS

> **Theorem 2.3.1 Euler's Theorem**
>
> Let a and n be two co-prime integers, then
> $$a^{\varphi(n)} \equiv 1 \pmod{n} \qquad (2.3)$$

When n is a prime, we have $\varphi(n) = n - 1$ and *Equation 2.3* becomes *(2.2)*.

Similar to Fermat's little theorem, Euler's theorem can also be used as the *positive one technique*. Let's consider an example.

Example 2.3.1

Calculate $3^{64} \pmod{67}$.

Solution

By Euler's theorem, we have $3^{66} \equiv 1 \pmod{67}$. Hence, we have $3^{64} \cdot 3^2 \equiv 1 \pmod{67}$. This means that the desired answer is just the multiplicative inverse of 9 modulo 67 which is $\boxed{15}$.

Done.

One application of Euler's theorem is to compute multiplicative inverse. When a and n are two co-prime integers, then
$$a^{-1} \equiv a^{\varphi(n)-1} \pmod{n}$$

The converse of Euler's theorem is also true.

> **Theorem 2.3.2 Converse to Euler's Theorem**
>
> If $a^{\varphi(n)} \equiv 1 \pmod{n}$ holds where a and n be two positive integers, then a and n are co-prime.

Euler's theorem can be a powerful tool to compute a^b (mod n) when a and n are co-prime. To tackle such a problem, it is sufficient to first compute $c \equiv b$ (mod $\varphi(n)$). Then we will have

$$a^b \equiv a^c \pmod{n}$$

2.4 Multiplicative Order

Given two co-prime integers a and n, the smallest positive integer k satisfying $a^k \equiv 1$ (mod n) is called the *multiplicative order* of a modulo n. It is usually written as $Ord_n(a)$ or $O_n(a)$.

By Euler's theorem, $Ord_n(a)$ cannot be larger than $\varphi(n)$. It can also be shown that $Ord_n(a)$ must divide $\varphi(n)$. Therefore, it is sufficient to check all the divisors of $\varphi(n)$ in order to determine $Ord_n(a)$.

Let's review an example.

Example 2.4.1

Find the multiplicative order of 2 modulo 17.

Solution

Let's examine all the divisors of $\varphi(17) = 16$. Because $2^8 \equiv 1$ (mod 17), but $2^4 \not\equiv 1$ (mod 17), therefore the answer is $\boxed{8}$.

Done.

2.5 Wilson's Theorem

When a MOD expression involves factorial, the Wilson' theorem may be relevant.

> **Theorem 2.5.1 Wilson's Theorem**
>
> An positive integer n is a prime if and only if
> $$(n-1)! \equiv -1 \pmod{n}$$

Chapter 3

MOD Equation

3.1 MOD Equation Basics

Let all the coefficients of a polynomial be integers
$$f(x) = a_n x^n + a_{n-1} x^{n-1} + \cdots + a_1 x + a_0$$

Then, *(3.1)* below is called a modular equation with respect to variable x:
$$f(x) \equiv 0 \pmod{m} \tag{3.1}$$

In many cases, a MOD equation can be simplified.

Example 3.1.1

Simplify following MOD equation
$$f(x) = 6x^3 + 5x + 2 \equiv 0 \pmod{3}$$

Solution

Because $6x^3 \equiv 0 \pmod{3}$ and $5x \equiv 2x \pmod{3}$, therefore
$$f(x) \equiv 2x + 2 \equiv 0 \pmod{3}$$

CHAPTER 3. MOD EQUATION

Done.

> **Theorem 3.1.1 Equivalent Modular Equations**
>
> Let $f(x)$, $g(x)$, and $h(x)$ be three polynomials with integer coefficients. If $f(x) = g(x)h(x) + r(x)$ and $h(x) \equiv 0 \pmod{m}$, then $f(x) \equiv 0 \pmod{m}$ and $r(x) \equiv 0 \pmod{m}$ are equivalent and have same solutions.

Let's consider an example.

Example 3.1.2

Simplify this modular equation:

$$f(x) = 3x^{15} - x^{13} - x^{12} - x^{11} - 3x^5 + 6x^3 - 2x^2 + 2x - 11 \equiv 0 \pmod{11}$$

Solution

Because 11 is a prime number, we conclude $x^{11} - x \equiv 0 \pmod{11}$ must be an identity by Fermat's little theorem. Dividing $f(x)$ by $(x^{11} - x)$ leads to

$$\begin{aligned} & 3x^{15} - x^{13} - x^{12} - x^{11} - 3x^5 + 6x^3 - 2x^2 + 2x - 11 \\ &= (x^{11} - x)(3x^4 - x^2 - x - 1) + (5x^3 - 3x^2 + x - 11) \\ \therefore \quad f(x) &\equiv 5x^3 - 3x^2 + x - 11 \\ &\equiv 5x^3 - 3x^2 + x \equiv 0 \pmod{11} \end{aligned}$$

Done.

A "brutal force" method to solve a MOD equation $f(x) \equiv 0 \pmod{m}$ is to try all the numbers in a complete residue system modulo m, such as $\{0, 1, 2, \cdots, m-1\}$. If one element of the residue system, let it be k, satisfies $f(k) \equiv 0 \pmod{m}$, then $x \equiv k \pmod{m}$ is one solution. There may be multiple solutions.

Not all MOD equations are solvable. In addition to using the brutal force to show no solution exists, there are several additional techniques available. One is to show that the two sides of the congruence relation do not share same residue. *Example 1.6.1* on *page 14* offers a good example. Another way is to utilize the following theorem.

> **Theorem 3.1.2 Necessary Condition**
>
> For $f(x) \equiv 0 \pmod{m}$ to be solvable, $f(x) \equiv 0 \pmod{d}$ must be solvable where d is any divisor of m.

Here is an example.

Example 3.1.3

Solve this modular equation:

$$f(x) = 4x^2 + 27x - 9 \equiv 0 \pmod{15}$$

Solutions is given in the practice.

3.2 Solving $ax \equiv b \pmod{m}$

If $ax \equiv b \pmod{m}$ is solvable, then there exists an integer y such that

$$ax - b = my \implies ax - my = b$$

This means that solving $ax \equiv b \pmod{m}$ is equivalent to solving the following linear indeterminate equation:

$$ax + by = c \qquad (3.2)$$

The standard way to solve *(3.2)* is the Euclidean method. It is discussed in the book *Number Theory - MOD and Indeterminate*

CHAPTER 3. MOD EQUATION

Equation in the Math All Star Series.

3.3 Chinese Remainder Theorem (CRT)

CRT relates to a system of congruent relations.

Theorem 3.3.1 Chinese Remainder Theorem

Let integers $m_1, m_2, \cdots, m_k$ be pair-wise co-prime and $a_1, a_2, \cdots, a_k$ be any integers, then the system

$$\begin{cases} x \equiv a_1 \pmod{m_1} \\ x \equiv a_2 \pmod{m_2} \\ \cdots \\ x \equiv a_k \pmod{m_k} \end{cases}$$

has a unique solution

$$x \equiv \sum_{i=1}^{k} a_i b_i b_i' \pmod{M}$$

where $M = m_1 m_2 \cdots m_k$, $b_i = M/m_i$, and $b_i' = b_i^{-1} \pmod{m_i}$.

It is worth noting that some special systems can be solved in an easier way. Let's consider an example.

Example 3.3.1

Find the smallest positive integer n such that

$$\begin{cases} n \equiv 1 \pmod{3} \\ n \equiv 3 \pmod{5} \\ n \equiv 5 \pmod{7} \end{cases}$$

Solution

Because 3, 5, and 7 are pair-wise co-prime, we can apply CRT

CHAPTER 3. MOD EQUATION

to solve this system.

$$\begin{aligned}
M &= 3 \times 5 \times 7 = 105 \\
b_1 &= 105 \div 3 = 35 &\implies b_1^{-1} &\equiv 35^{-1} \equiv 2 \pmod{3} \\
b_2 &= 105 \div 5 = 21 &\implies b_2^{-1} &\equiv 21^{-1} \equiv 1 \pmod{5} \\
b_3 &= 105 \div 7 = 15 &\implies b_3^{-1} &\equiv 15^{-1} \equiv 1 \pmod{7}
\end{aligned}$$

Therefore, by CRT, we have the solution as

$$n \equiv 1 \times 35 \times 2 + 3 \times 21 \times 1 + 5 \times 15 \times 1 \equiv 103 \pmod{105}$$

The smallest positive integer satisfying this relation is $\boxed{103}$.

Meanwhile, this problem can also be solved without employing CRT. We note that all the residues are 2 less than the modulo. Therefore the answer must be 2 less the least common divisor of 3, 5, and 7 which is $\boxed{103}$.

Done.

CHAPTER 3. MOD EQUATION

Part II

Practice

Practice 1

Let $a > b > c$ be three positive integers. If their remainders are 2, 7, and 9 respectively when being divided by 11. Find the remainder when $(a+b+c)(a-b)(b-c)$ is divided by 11.

(ref: 122)

Practice 2

Let four positive integers a, b, c, and d satisfy $a+b+c+d = 2019$. Prove $(a^3 + b^3 + c^3 + d^3)$ cannot be an even number.

(ref: 1120)

Practice 3

Prove that $7 \mid 8^n - 1$ for $n \geq 1$.

(ref: 3191)

Practice 4

Show that $5 \mid 4^{2n} - 1$ for $n \geq 1$.

(ref: 3192)

Practice 5

Prove that $15 \mid 4^{2n} - 1$ for $n \geq 1$.

(ref: 3194)

Practice 6

When Ringo places his marbles into bags with 6 marbles per bag, he has 4 marbles left over. When Paul does the same with his marbles, he has 3 marbles left over. Ringo and Paul pool their marbles and place them into as many bags as possible, with 6 marbles per bag. How many marbles will be left over?

(ref: 1408 - AMC10)

Practice 7

Show that there exists an infinite number of integers in the form of $(2^n + 27)$ which are multiples of 7.

(ref: 4253)

Practice 8

Let integer a, b, and c satisfy $a+b+c=0$, prove $|a^{1999}+b^{1999}+c^{1999}|$ is a composite number.

(ref: 310)

Practice 9

If 2016 consecutive integers are added together, where the 999th number in the sequence is $1,244,584$, what is the remainder when this sum is divided by 6?

(ref: 3070)

Practice 10

Let integers x, y, z satisfy
$$(x-y)(y-z)(z-x) = x+y+z$$
Show that $27 \mid (x+y+z)$

(ref: 4203)

Practice 11

Suppose integers a and b satisfy $ab \equiv -1 \pmod{24}$. Prove $(a+b)$ must be a multiple of 24.

(ref: 4215)

Practice 12

Find the largest integer x such that for any positive integer y, the number $(7^y + 12y - 1)$ is always a multiple of x.

(ref: 4244)

Practice 13

What is the last digit of 9^{2019}?

(ref: 272)

Practice 14

What are the last two digits of 8^{88}?

(ref: 273)

Practice 15

What is the tens digit of 7^{2019}?

(ref: 1171)

Practice 16

Find the remainder when $3^{2019} + 4^{2019}$ is divided by 5?

(ref: 274)

Practice 17

Find the remainder when $9 \times 99 \times 999 \times \cdots \times \underbrace{99\cdots 9}_{999}$ is divided by 1000.

(ref: 280 - AIME)

Practice 18

An integer N is selected at random in the range $1 \leq N \leq 2020$. What is the probability that the remainder when N^{16} is divided by 5 is 1?

(ref: 3743 - AMC10)

Practice 19

Determine the units digit of the sum $0! + 1! + 2! + \cdots + n! + \cdots + 20!$?

(ref: 2654 - BCML)

Practice 20

What are the last two digits in the sum of the factorials of the first 100 positive integers?

(ref: 1117)

Practice 21

Find the largest positive integer n such that $(3^{1024} - 1)$ is divisible by 2^n.

(ref: 2811)

Practice 22

Let n be any positive integer, show that

$$(5n+1)(5n+2)(5n+3)(5n+4) \equiv -1 \pmod{25}$$

(ref: 4160)

Practice 23

Let N be the product of four consecutive odd numbers. Show that $N \equiv 1 \pmod 8$.

(ref: 4247)

Practice 24

Let the product of all odd positive integer not greater than 2019 be 2019!!. Find the last three digits of 2019!!.

(ref: 4248)

Practice 25

Let n^2 be a square. Show that $n^2 \equiv 0, 1 \pmod{3}$.

(ref: 4206)

Practice 26

Let k be a positive integer, show that $(4k + 3)$ cannot be a square number.

(ref: 2360)

Practice 27

Let n^2 be a square number. Show that $n^2 \equiv 0, \pm 1 \pmod{5}$.

(ref: 4156)

Practice 28

Show that if n^2 is a square number, then $n^2 \equiv 0, 1, 4, 9 \pmod{16}$.

(ref: 4144)

Practice 29

Show that the difference of two squares of odd numbers must be a multiple of 8.

(ref: 2250)

Practice 30

How many terms in this sequence are squares?

$$1, 11, 111, 1111, \cdots$$

(ref: 4147)

Practice 31

How many terms in this sequence are squares?

$$4, 44, 444, 4444, \cdots$$

(ref: 4148)

Practice 32

How many numbers in this series are squares?

$$1, 14, 144, 1444, 14444, \cdots$$

(ref: 2361)

Practice 33

Find the number of integer pairs (x, y) such that $x^2 + y^2 = 2019$.

(ref: 4140)

Practice 34

Solve this equation in integers: $x_1^4 + x_2^4 + \cdots + x_{14}^4 = 9999$.

(ref: 4214)

Practice 35

What is the smallest positive integer greater than 5 which leaves a remainder of 5 when divided by each of 6, 7, 8, and 9?

(ref: 2645 - BCML)

Practice 36

A box contains gold coins. If the coins are equally divided among six people, four coins are left over. If the coins are equally divided among five people, three coins are left over. If the box holds the smallest number of coins that meets these two conditions, how many coins are left when equally divided among seven people?

(ref: 991 - AMC8)

Practice 37

Three runners start running simultaneously from the same point on a 500-meter circular track. They each run clockwise around the course maintaining constant speeds of 4.4, 4.8, and 5.0 meters per second. The runners stop once they are all together again somewhere on the circular course. How many seconds do the runners run?

(ref: 1395 - AMC10)

Practice 38

Show that there is at least one Friday 13th in any year, including any leap year.

(ref: 4252)

Practice 39

The number 2^{29} is a nine-digit number whose digits are all distinct. Which digit of 0 to 9 does not appear?

(ref: 311)

Practice 40

Select nine different digits from 0 to 9 to form a two-digit number, a three-digit number and a four-digit number. The sum of these three numbers is 2017. Which digit is not selected?

(ref: 4221)

Practice 41

Find the multiplicative order of 3 modulo 17.

(ref: 4178)

Practice 42

Find the multiplicative order of 5 modulo 19.

(ref: 4179)

Practice 43

Find the multiplicative order of 2 modulo 125.

(ref: 4191)

Practice 44

Let n be an odd integer greater than 1, then n is the multiplicative order of 2 modulo $(2^n - 1)$.

(ref: 4182)

Practice 45

Let p be an odd prime, and integer a has multiplicative order of $2k$ modulo p, then $a^k \equiv -1 \pmod{p}$.

(ref: 4181)

Practice 46

Show that if integer a has multiplicative order of hk modulo n, then a^h has order of k modulo n.

(ref: 4180)

Practice 47

Let $n > 4$ be a composite number. Show that $(n-1)! \equiv 0 \pmod{n}$.

(ref: 4196)

Practice 48

Let n be an integer greater than 1. If none of $1!, 2!, \cdots, n!$ has the same remainder when being divided by n, show that n is a prime.

(ref: 4202)

Practice 49

Show that for any positive integer k, it always holds that $10^k \equiv 4 \pmod{6}$.

(ref: 4224)

Practice 50

Show that for any integer x, the number $\left(\frac{x^5}{5} + \frac{x^3}{3} + \frac{7x}{15}\right)$ is an integer.

(ref: 4262)

Practice 51

Show that $(2^{1194} + 1)$ is a multiple of 65.

(ref: 4167)

Practice 52

Let $k = 2008^2 + 2^{2008}$. What is the units digit of $k^2 + 2^k$?

(ref: 1608 - AMC10)

Practice 53

Compute $9^{50} \pmod{1000}$.

(ref: 4216)

Practice 54

Find the last three digits of $9 + 9^2 + 9^3 + \cdots + 9^{2000}$.

(ref: 4217)

Practice 55

Find the remainder when $10^{10} + 10^{100} + 10^{1000} + \cdots + 10^{\overbrace{10\cdots 0}^{2018}}$ is divided by 7.

(ref: 4225)

Practice 56

Given 30! ends with some zeros, what is the digit that immediately precedes these zeros?

(ref: 4173)

Practice 57

The two-digit integers from 19 to 92 are written consecutively to form the large integer

$$N = 192021 \cdots 909192$$

Suppose that the 3^k is the highest power of 3 that is a factor of N. What is k.

(ref: 4174 - AMC12)

Practice 58

Let S be the sum of squares of 10 consecutive positive integers. Show S cannot be a square.

(ref: 4250)

Practice 59

What is the remainder when $(8888^{2222} + 7777^{3333})$ is divided by 37?

(ref: 167)

Practice 60

Compute 3^{2018} mod 17.

(ref: 4161)

Practice 61

Compute $50^{250} \pmod{83}$.

(ref: 2740)

Practice 62

Compute 20! (mod 23).

(ref: 4169)

Practice 63

Let $N = 7 \times 8 \times 9 \times 15 \times 16 \times 17 \times 23 \times 24 \times 25 \times 43$. Compute $N \pmod{11}$.

(ref: 4218)

Practice 64

Let p is an odd prime, compute $1^{p-1} + 2^{p-1} + 3^{p-1} + \cdots + (p-1)^{p-1} \pmod{p}$.

(ref: 4228)

Practice 65

Let p is an odd prime, compute $1^p + 2^p + 3^p + \cdots + (p-1)^p \pmod{p}$.

(ref: 4229)

Practice 66

Find the smallest positive integer n so that $107n$ has the same last two digits as n.

(ref: 2805 - Harvard-MIT)

Practice 67

How many positive integers N, less than 2017, satisfy

$$N^{2016^{2016}} \equiv 1 \pmod{2017}$$

(ref: 4227)

Practice 68

Solve $x^{12} \equiv 3 \pmod{11}$.

(ref: 4164)

Practice 69

Solve this modular equation:
$$f(x) = 4x^2 + 27x - 9 \equiv 0 \pmod{15}$$

(ref: 4240)

Practice 70

Compute $3^{2017} \pmod{1000}$.

(ref: 4241)

Practice 71

Solve the system of congruence
$$\begin{cases} x \equiv 1 \pmod{3} \\ x \equiv 2 \pmod{5} \\ x \equiv 3 \pmod{7} \end{cases}$$

(ref: 4197)

Practice 72

Show that if there exist integer x, y, and z such that $3^x + 4^y = 5^z$, then both x and z must be even.

(ref: 4185)

Practice 73

Solve the congruent system: $4x \equiv 2 \pmod{6}$ and $3x \equiv 5 \pmod{8}$.

(ref: 4199)

Practice 74

Show that $2x^2 - 5y^2 = 7$ has no integer solution.

(ref: 125)

Practice 75

Let m be the least positive integer divisible by 17 whose digits sum is 17. Find m.

(ref: 70 - AIME)

Practice 76

Let $f(n)$ denote the sum of the digits of n. Find $f(f(f(4444^{4444})))$.

(ref: 2216 - IMO)

Practice 77

What is the last digit of $17^{17^{17^{17}}}$?

(ref: 2540 - PUMaC)

Practice 78

Let $N = 4568^{7777}$, a be the sum of digits in N, b be the sum of digits in a, and c be the sum of digits in b. Find c.

(ref: 4261)

Practice 79

Let $\mathbb{S}$ be a set containing all the integers created by digits 1, 2, $\cdots$, 7. Each digit can be used once and only once. Show that no element in $\mathbb{S}$ is a multiple of the other.

(ref: 4208)

Practice 80

Let p be a prime and integer a is co-prime to p, show that

$$a^{p(p-1)} \equiv 1 \pmod{p^2}$$

(ref: 4170)

Practice 81

Let p and q be two distinct primes, and integer a is co-prime to both p and q, show

$$a^{(p-1)(q-1)} \equiv 1 \pmod{pq}$$

(ref: 4171)

Practice 82

Show that two positive integers m and n are co-prime if and only if $\varphi(mn) = \varphi(m)\varphi(n)$.

(ref: 4195)

Practice 83

Let p be a prime and k be a positive integer less than p. Show that $C_p^k \equiv 0 \pmod{p}$.

(ref: 4232)

Practice 84

Show that $\varphi(n) = n/4$ is impossible to hold.

(ref: 4220)

Practice 85

Let $P(x)$ be a polynomial with integer coefficients satisfying that both $P(0)$ and $P(1)$ are odd. Show that $P(x)$ has no integer zeros.

(ref: 2841)

Practice 86

Does there exist a polynomial $P(x)$ such that $P(1) = 2015$ and $P(2015) = 2016$?

(ref: 3971)

Practice 87

Find 8 prime numbers, not necessarily distinct such that the sum of the squares of these numbers is 992 less than 4 times of the product of these numbers.

(ref: 2822)

Practice 88

Show that for any positive integer n, $\varphi(2^n - 1)$ is a multiple of n where $\varphi(n)$ is Euler's totient function.

(ref: 4183)

Practice 89

Let p be an odd prime divisor of integer $(n^4 + 1)$. Show that $p \equiv 1 \pmod{8}$.

(ref: 4184)

Practice 90

Let n be a positive odd integer. Show that at least one of the following numbers is a multiple of n.

$$2 - 1, 2^2 - 1, \cdots, 2^{n-1} - 1$$

(ref: 4204)

Practice 91

Show that from any given m integers, it is always possible to select one or more integers such that their sum is a multiple of m.

(ref: 4223)

Practice 92

Let sequence $\{x_n\}$ satisfy the relation $x_{n+2} = x_{n+1} + 2x_n$ for $n \geq 1$ where $x_1 = 1$ and $x_2 = 3$.

Let sequence $\{y_n\}$ satisfy the relation $y_{n+2} = 2y_{n+1} + 3y_n$ for $n \geq 1$ where $y_1 = 7$ and $y_2 = 17$.

Show that these two sequences do not share any common term.

(ref: 4209)

Practice 93

Show that if n is an integer greater than 1, then $(2^n - 1)$ is not divisible by n.

(ref: 3624 - Putnam)

Practice 94

Let p be an odd prime divisor of number $(a^2 + 1)$ where a is an integer. Show that $p \equiv 1 \pmod{4}$.

(ref: 3870)

Practice 95

Let a and b be two positive integers such that both of them can be written as a sum of two squares. Show that their product can be written as a sum of two squares in two ways.

(ref: 4189)

Practice 96

Let $\{a_1, a_2, \cdots, a_{2n+1}\}$ be a set of integers such that after removing any element, the remaining ones can always be equally divided into two groups with equal sum. Show that all these a_i, $(1 \leq i \leq 2n+1)$ are equal.

(ref: 4255 - Putnam)

Practice 97

Let x and y be two integers and p be a prime. Show that

$$(x+y)^p \equiv x^p + y^p \pmod{p}$$

(ref: 4233)

Practice 98

Let n be a positive integer and k be an odd positive integer, show $k^{2^n} \equiv 1 \pmod{2^{n+2}}$.

(ref: 4243)

Practice 99

Let m and n be positive integers, m be odd, and $(m, 2^n-1) = 1$.
Show that $\sum_{k=1}^{m} k^n$ is a multiple of m.

(ref: 4268)

Practice 100

The number obtained from the last two non-zero digits of 90! is equal to n. What is n?

(ref: 1508 - AMC10)

Practice 101

Find all ordered integer pairs (x, y) such that $x^3 + y^3 = 2019$.

(ref: 4249)

Practice 102

Find the least non-negative residue of 70! (mod 5183).

(ref: 2739)

Practice 103

What is the smallest positive integer n such that $20 \equiv n^{15}$ (mod 29)?

(ref: 2614 - PUMaC)

Practice 104

Find one solution to $x^7 \equiv 3 \pmod{11}$.

(ref: 4166)

Practice 105

Solve
$$\begin{cases} x &\equiv 2 \pmod 3 \\ x &\equiv 2 \pmod 5 \\ x &\equiv -3 \pmod 7 \\ x &\equiv -2 \pmod{13} \end{cases}$$

(ref: 2639)

Practice 106

Solve
$$\begin{cases} 4x &\equiv 14 \pmod{15} \\ 9x &\equiv 11 \pmod{20} \end{cases}$$

(ref: 2662)

Practice 107

Let n be a positive integer and function $S_1(n)$ return the square of the sum of n's digits. Additionally, let $S_{k+1}(n) = S_1(S_k(n))$, where k is a positive integer. Find the value of $S_{1991}(2^{1990})$.

(ref: 4245 - China)

Practice 108

Find all the integer pairs (x, y) such that $x^3 = 2^y + 15$.

(ref: 4254)

Practice 109

Show that $x^5 \equiv 3 \pmod{11}$ is not solvable.

(ref: 4165)

Practice 110

Solve the following relation in integers:
$$x^2 + a^2 = (x+1)^2 + b^2 = (x+2)^2 + c^2 = (x+3)^2 + d^2$$

(ref: 4226)

Practice 111

Show that there are infinite many composite numbers in the sequence
$$1, 31, 331, 3331, 33331, \cdots$$

(ref: 4212)

Practice 112

Let n be a positive integer. Show that there exist n consecutive integers each of which contains a divisor who is a square number greater than 1.

(ref: 4213)

Practice 113

Let a, b, and x_0 all be positive integers. Sequence $\{x_n\}$ is defined as $x_{n+1} = ax_n + b$ where $n \geq 1$. Show that $x_1, x_2, \cdots$ cannot be all prime.

(ref: 4257)

Practice 114

Let n be an integer which is divisible by neither 2 nor 5. Show that n must be divisible by a number whose digits are all 1.

(ref: 4260)

Practice 115

Find the smallest integer N such that $\varphi(n) \geq 5$ holds for all integer $n \geq N$.

(ref: 4194)

Practice 116

Find all prime number p such that both $(4p^2 + 1)$ and $(6p^2 + 1)$ are prime numbers.

(ref: 174 - Poland)

Practice 117

Ms. Math's kindergarten class has 16 registered students. The classroom has a very large number, N, of play blocks which satisfies the conditions:

- If 16, 15, or 14 students are present in the class, then in each case all the blocks can be distributed in equal numbers to each student, and

- There are three integers $0 < x < y < z < 14$ such that when x, y, or z students are present and the blocks are distributed in equal numbers to each student, there are exactly three blocks left over.

Find the sum of the distinct prime divisors of the least possible value of N satisfying the above conditions.

(ref: 190 - AIME)

Practice 118

Let S be the set of all perfect squares whose rightmost three digits in base 10 are 256. Let T be the set of all numbers of the form $\frac{x-256}{1000}$, where x is in S. In other words, T is the set of numbers when the last three digits of each number in S are truncated. Find the remainder when the tenth smallest element of T is divided by 1000.

(ref: 219 - AIME)

Practice 119

For a positive integer p, define the positive integer n to be p-safe if n differs in absolute value by more than 2 from all multiples of p. For example, the set of 10-safe numbers is $\{3, 4, 5, 6, 7, 13, 14, 15, 16, 17, 23, \ldots\}$. Find the number of positive integers less than or equal to $10,000$ which are simultaneously 7-safe, 11-safe, and 13-safe.

(ref: 236 - AIME)

Practice 120

Let R be the set of all possible remainders when a number of the form 2^n, where n is a non-negative integer, is divided by 1000. Let S be the sum of the elements in R. Find the remainder when S is divided by 1000.

(ref: 250 - AIME)

Practice 121

The number 2017 is prime. Let $S = \sum_{k=0}^{62} \binom{2014}{k}$. What is the remainder when S is divided by 2017?

(ref: 474 - AMC12)

Practice 122

Seven students count from 1 to 1000 as follows:

- Alice says all the numbers, except she skips the middle number in each consecutive group of three numbers. That is, Alice says 1, 3, 4, 6, 7, 9, . . ., 997, 999, 1000.

- Barbara says all of the numbers that Alice doesn't say, except she also skips the middle number in each consecutive group of three numbers.

- Candice says all of the numbers that neither Alice nor Barbara says, except she also skips the middle number in each consecutive group of three numbers.

- Debbie, Eliza, and Fatima say all of the numbers that none of the students with the first names beginning before theirs in the alphabet say, except each also skips the middle number in each of her consecutive groups of three numbers.

- Finally, George says the only number that no one else says.

What number does George say?

(ref: 1457 - AMC10)

Practice 123

N delegates attend a round-table meeting, where N is an even number. After a break, these delegates randomly pick a seat to sit down again to continue the meeting. Prove that there must exist two delegates so that the number of people sitting between them is the same before and after the break.

(ref: 2088)

Practice 124

Prove that if p and $(p^2 + 8)$ are prime, then $(p^3 + 8p + 2)$ is prime.

(ref: 2217)

Practice 125

Given that there are 24 primes between 3 and 100, inclusive, what is the number of ordered pairs (p, a) with p prime, $3 \leq p < 100$, and $1 \leq a < p$ such that the sum $a + a^2 + a^3 + \cdots + a^{(p-2)!}$ is not divisible by p?

(ref: 2615 - PUMaC)

Practice 126

If for any integer $k \neq 27$ and $\left(a - k^{2015}\right)$ is divisible by $(27 - k)$, what is the last two digits of a?

(ref: 2621)

Practice 127

A positive integer n is said to be good if there exists a perfect square whose sum of digits in base 10 is equal to n. For instance, 13 is good because $7^2 = 49$ and $4 + 9 = 13$. How many good numbers are among $1, 2, 3, \cdots, 2007$?

(ref: 2816)

Practice 128

Let x be an integer and p is a prime divisor of $(x^6 + x^5 + \cdots + 1)$. Show that $p = 7$ or $p \equiv 1 \pmod{7}$.

(ref: 3795)

Practice 129

Let p be an odd prime number. For positive integer k satisfying $1 \leq k \leq p-1$, the number of divisors of $kp+1$ between k and p exclusive is a_k. Find the value of $a_1 + a_2 + \ldots + a_{p-1}$.

(ref: 3841 - Japan)

Practice 130

Compute $\underbrace{3^{3^{3^{\cdots^3}}}}_{2012 \ times}$ $\pmod{100}$.

(ref: 4162)

Practice 131

How many positive integers not exceeding 100 are there such that the value of $(3^x - x^2)$ is a multiple of 5?

(ref: 4176)

Practice 132

Let $\mathbb{S}$ be the set of integers between 1 and 2^{40} that contain two 1s when written in base 2. What is the probability that a random integer from $\mathbb{S}$ is divisible by 9?

(ref: 4177 - AIME)

Practice 133

(Thue's theorem) Let p be a prime. Show that for any integer a such that $p \nmid a$, there exist positive integers x, y not exceeding $\lfloor\sqrt{p}\rfloor$ satisfying $ax \equiv y \pmod{p}$ or $ax \equiv -y \pmod{p}$.

(ref: 4186)

Practice 134

Let p be a prime. Show that there exist infinitely many positive integer n such that $p \mid (2^n - n)$.

(ref: 4205)

Practice 135

Let p be a prime number and $\lfloor x \rfloor$ denote the largest integer not exceeding real number x. Show that

$$C_n^p \equiv \left\lfloor \frac{n}{p} \right\rfloor \pmod{p}$$

(ref: 4222)

Practice 136

Let integer $N = \left\lfloor (\sqrt{29} + \sqrt{21})^{2020} \right\rfloor$ where $\lfloor x \rfloor$ denotes the largest integer not exceeding x. Find the last two digits of N.

(ref: 4242)

Practice 137

Let m and n be two positive integers, find the minimal value of $|\,12^m - 5^n\,|$.

(ref: 4246)

Practice 138

Determine all positive integer n such that the following equation is solvable in integers:

$$x^n + (2+x)^n + (2-x)^n = 0$$

(ref: 4258)

Practice 139

Let sequence $\{a_n\}$ be $a_n = 2^n + 3^n + 6^n - 1$ where $n \geq 1$. Find the sum of all positive integers which are co-prime to all the a_n.

(ref: 4269)

Practice 140

Let p be a prime and

$$\frac{a}{b} = \frac{1}{1^2} + \frac{1}{2^2} + \cdots + \frac{1}{(p-1)^2}$$

where a and b are two co-prime positive integers. Show that $p \mid a$.

(ref: 4219)

Practice 141

Let n be an odd integer greater than 3, and $\mathbb{S} = \{0, 1, \cdots, n-1\}$. Show that after removing any element from $\mathbb{S}$, it is always possible to equally divide the remaining elements in $\mathbb{S}$ into two groups such that their sum are congruent modulo n.

(ref: 4210)

Practice 142

Let n be a positive integer not less than 4. Show that there exists a polynomial with integral coefficients

$$f(x) = x^n + a_{n-1}x^{n-1} + a_{n-2}x^{n-2} + \cdots + a_1 x + a_0$$

such that for any positive integer m and any $k \geq 2$ distinct integers $r_1, r_2, \cdots, r_k$, it always hold that $f(m) \neq f(r_1)f(r_2)\cdots f(r_k)$.

(ref: 4211)

Practice 143

(Fermat's little theorem) Show that $a^p \equiv a \pmod{p}$ holds if p is a prime.

(ref: 4235)

Practice 144

An integer in the form of $F_n = 2^{2^n} + 1$ where integer $n \geq 1$ is called a Fermat's number. Let d_n be any divisor of F_n. Show that $d_n \equiv 1 \pmod{2^{n+1}}$.

(ref: 4237)

Practice 145

Assume positive integer $n > 1$ satisfies $n \mid (2^n + 1)$, prove n is a multiple of 3.

(ref: 4238)

Practice 146

Let p be an odd prime, and $n = \frac{2^{2p}-1}{3}$ in an integer. Prove $2^{n-1} \equiv 1 \pmod{n}$.

(ref: 4239)

Practice 147

Show that there exists an infinite number of squares in the form of $(n \cdot 2^k - 7)$ where n and k are both positive integers.

(ref: 4251)

Practice 148

Show that the number $(2n^{3k} + 4n^k + 10)$ cannot be a product of consecutive integers for any positive integers n and k.

(ref: 4256)

Practice 149

Let integers $l > m > n$ be the side lengths of a triangle satisfying $\left\{\frac{3^l}{10^4}\right\} = \left\{\frac{3^m}{10^4}\right\} = \left\{\frac{3^n}{10^4}\right\}$ where function $\{x\}$ returns the decimal part of real number x. Find the least possible value of this triangle's perimeter.

(ref: 4259)

Practice 150

Solve $x^{22} + x^{11} \equiv 2 \pmod{11}$.

(ref: 4168)

Practice 151

Let sequence $g(n)$ satisfy $g(1) = 0, g(2) = 1, g(n+2) = g(n+1) + g(n) + 1$ where $n \geq 1$. Show that if n is a prime greater than 5, then $n \mid g(n)[g(n) + 1]$.

(ref: 2699 - IMO)

Practice 152

Let p be an odd prime. Show that

$$\sum_{j=0}^{p} C_p^j C_{p+j}^j \equiv 2^p + 1 \pmod{p^2}$$

(ref: 4172 - Putnam)

Practice 153

Show that if the equation $a^2 + 1 \equiv 0 \pmod{p}$ is solvable for some a, then p can be represented as a sum of two squares.

(ref: 4187)

Practice 154

Show that a prime $p > 2$ is a sum of two squares if and only if $p \equiv 1 \pmod{4}$.

(ref: 4188)

Practice 155

(Two Squares Theorem) Show that a positive integer n is a sum of two squares if and only if each prime factor p of n such that $p \equiv 3 \pmod{4}$ occurs to an even power in the prime factorization of n.

(ref: 4190)

Part III

Solution

Practice 1

Let $a > b > c$ be three positive integers. If their remainders are 2, 7, and 9 respectively when being divided by 11. Find the remainder when $(a+b+c)(a-b)(b-c)$ is divided by 11.

(ref: 122)

$(a+b+c)(a-b)(b-c) \equiv (2+7+9)(2-7)(7-9) \equiv 180 \equiv \boxed{4} \pmod{11}$

Practice 2

Let four positive integers a, b, c, and d satisfy $a+b+c+d = 2019$. Prove $(a^3 + b^3 + c^3 + d^3)$ cannot be an even number.

(ref: 1120)

It is easy to show that regardless of integer n's parity, it always hold that $n^3 \equiv n \pmod 2$ because any power of n will not change odd even parity. Therefore,

$$a^3 + b^3 + c^3 + d^3 \equiv a+b+c+d \equiv 2019 \equiv 1 \pmod 2$$

Therefore, it is odd.

Practice 3

Prove that $7 \mid 8^n - 1$ for $n \geq 1$.

(ref: 3191)

Because $8^n - 1 \equiv 1^n - 1 \equiv 0 \pmod 7$, therefore the conclusion holds.

Practice 4

Show that $5 \mid 4^{2n} - 1$ for $n \geq 1$.

(ref: 3192)

Because $4^{2n} - 1 \equiv (-1)^{2n} - 1 \equiv 1^n - 1 \equiv 0 \pmod{5}$, therefore the claim holds.

Practice 5

Prove that $15 \mid 4^{2n} - 1$ for $n \geq 1$.

(ref: 3194)

Because $4^{2n} - 1 \equiv 16^n - 1 \equiv 0 \pmod{15}$, therefore the conclusion holds.

Practice 6

When Ringo places his marbles into bags with 6 marbles per bag, he has 4 marbles left over. When Paul does the same with his marbles, he has 3 marbles left over. Ringo and Paul pool their marbles and place them into as many bags as possible, with 6 marbles per bag. How many marbles will be left over?

(ref: 1408 - AMC10)

The remainder will be $4 + 3 \equiv 7 \equiv \boxed{1} \pmod{6}$.

Practice 7

Show that there exists an infinite number of integers in the form of $(2^n + 27)$ which are multiples of 7.

(ref: 4253)

Note that $8 \equiv 1 \pmod 7$. Therefore $2^{3k} \equiv 1 \pmod 7$ holds for any positive integer k. It follows that

$$2^{3k} + 27 \equiv 1 + 27 \equiv 0 \pmod 7$$

This means when n is a multiple of 3, the number $(2^n + 27)$ will be a multiple of 7. Clearly, there is an infinite number of such n.

Practice 8

Let integer a, b, and c satisfy $a+b+c = 0$, prove $|a^{1999} + b^{1999} + c^{1999}|$ is a composite number.

(ref: 310)

Let $d = a^{1999} + b^{1999} + c^{1999}$, we are going to show that d is a multiple of 6 which means $|d|$ is a composite.

Firstly, d is a multiple of 2 because

$$d \equiv a^{1999} + b^{1999} + c^{1999} \equiv a+b+c \equiv 0 \pmod 2$$

Next, by Fermat's little theorem, we have $x^3 \equiv x \pmod 3$. This conclusion can also be reasoned by factorizing $x^3 - x = (x-1)x(x+1)$ which is a product of three consecutive integers. One of them must be a multiple of 3. Hence $x^3 - x \equiv 0 \pmod 3$. It follows that

$$\begin{aligned} d &\equiv a \cdot a^{1998} + b \cdot b^{1998} + c \cdot c^{1998} \\ &\equiv a \cdot a^{666} + b \cdot b^{666} + c \cdot c^{666} \\ &\equiv a \cdot a^{222} + b \cdot b^{222} + c \cdot c^{222} \\ &\equiv a \cdot a^{74} + b \cdot b^{74} + c \cdot c^{74} \\ &\equiv a^{75} + b^{75} + c^{75} \\ &\equiv a + b + c \\ &\equiv 0 \pmod 3 \end{aligned}$$

This means d is a multiple of 3. Therefore, it must be a multiple of $2 \times 3 = 6$.

Practice 9

If 2016 consecutive integers are added together, where the 999th number in the sequence is $1,244,584$, what is the remainder when this sum is divided by 6?

(ref: 3070)

Because 2016 is a multiple of 6, therefore, the remainders of the sum of these numbers divides 6 will equal

$$(0+1+2+\cdots+5) \times 2016/6 \equiv \boxed{0} \pmod 6$$

Practice 10

Let integers x, y, z satisfy

$$(x-y)(y-z)(z-x) = x+y+z$$

Show that $27 \mid (x+y+z)$

(ref: 4203)

If the remainders of x, y, and z divided by 3 are all distinct, then we have

$$x+y+z \equiv 0+1+2 \equiv 0 \pmod 3$$

But if they are all distinct, then $(x-y)(y-z)(z-x) \not\equiv 0 \pmod 3$. Hence, this will contradict to the given condition. It follows that at least two of them are congruent modulo 3. By symmetry, let's assume $x \equiv y \pmod 3$.

Now we have

$$\begin{aligned}
& (x-y)(y-z)(z-x) \equiv 0 && \pmod 3 \\
\implies & x+y+z \equiv (x-y)(y-z)(z-x) \equiv 0 && \pmod 3 \\
\implies & z \equiv -(x+y) \equiv -2x \equiv x && \pmod 3 \\
\implies & x \equiv y \equiv z && \pmod 3
\end{aligned}$$

∴ $(x-y) \equiv (y-z) \equiv (z-x) \equiv 0 \pmod{3} \implies 27 \mid (x-y)(y-z)(z-x)$

Hence, we find $27 \mid (x+y+z)$ because $(x+y+z) = (x-y)(y-z)(z-x)$.

Practice 11

Suppose integers a and b satisfy $ab \equiv -1 \pmod{24}$. Prove $(a+b)$ must be a multiple of 24.

(ref: 4215)

The fact $ab \equiv -1 \pmod{24}$ means that $ab \equiv -1 \pmod{3}$, $ab \equiv -1 \pmod{8}$, and neither a nor b is zero.

Let's first consider $ab \equiv -1 \pmod{3}$. If $a \equiv \pm 1 \pmod{3}$, then $b \equiv \mp 1 \pmod{3}$. Either way, $(a+b) \equiv 0 \pmod{3}$.

Next, let's consider $ab \equiv -1 \pmod{8}$. It is clearly $a \pmod{8}$ can only be an odd number, i.e. $a \equiv \pm 1, \pm 3 \pmod{8}$.

- If $a \equiv \pm 1 \pmod{8}$, then $b \equiv \mp 1 \pmod{8}$. It follows that $(a+b) \equiv 0 \pmod{8}$.
- If $a \equiv \pm 3 \pmod{8}$, then $b \equiv \mp 3 \pmod{8}$. It follows that $(a+b) \equiv 0 \pmod{8}$.

Therefore, we always have $(a+b) \equiv 0 \pmod{8}$.

These two mean that $(a+b)$ must be a multiple of 24.

Practice 12

Find the largest integer x such that for any positive integer y, the number $(7^y + 12y - 1)$ is always a multiple of x.

(ref: 4244)

When $y = 1$, $7^y + 12y - 1 = 18$. Therefore, we must have $x \leq 18$. We are going to show that for any y, $18 \mid (7^y + 12y - 1)$. If so, the answer will be $\boxed{18}$.

Because $(7^y + 12y - 1)$ is obviously an even number, it is sufficient to show it is also a multiple of 9 in order to prove it is a multiple of 18.

- if $y \equiv 0 \pmod 3$, let it be $y = 3k$ where k is a positive integer. Then
$$7^y + 12y - 1 \equiv \left(7^3\right)^k + 12y - 1 \equiv 1^k + 36k - 1 \equiv 0 \pmod 9$$

- if $y \equiv 1 \pmod 3$, let it be $y = 3k + 1$ where k is a positive integer. Then
$$7^y + 12y - 1 \equiv 7 \cdot \left(7^3\right)^k + 12y - 1 \equiv 7 \cdot 1^k + 36k + 3 - 1 \equiv 0 \pmod 9$$

- if $y \equiv 2 \pmod 3$, let it be $y = 3k + 2$ where k is a positive integer. Then
$$7^y + 12y - 1 \equiv 7^2 \cdot \left(7^3\right)^k + 12y - 1 \equiv 7^2 \cdot 1^k + 36k + 6 - 1 \equiv 0 \pmod 9$$

Therefore, it is always a multiple of 9.

Practice 13

What is the last digit of 9^{2019}?

(ref: 272)

$$9^{2015} \equiv (-1)^{2019} \equiv -1 \equiv \boxed{9} \pmod{10}$$

Practice 14

What are the last two digits of 8^{88}?

(ref: 273)

$$\begin{aligned} 8^{88} &\equiv 2^{264} \\ &\equiv (2^{10})^{26} \times 2^4 \\ &\equiv 24^{26} \times 16 \\ &\equiv 76^{13} \times 16 \\ &\equiv 76 \times 16 \\ &\equiv \boxed{16} \pmod{100} \end{aligned}$$

Practice 15

What is the tens digit of 7^{2019}?

(ref: 1171)

We note that any power of 7^4 must end with 01. Therefore
$$7^{2019} = (7^4)^{504} \times 7^3 \equiv (01)^{504} \times 43 \equiv 43 \pmod{100}$$
Hence the answer is $\boxed{4}$.

Practice 16

Find the remainder when $3^{2019} + 4^{2019}$ is divided by 5?

(ref: 274)

$$3^{2019} + 4^{2019} \equiv (3^2)^{1009} \times 3 + (-1)^{2019} \equiv (-1)^{1009} \times 3 - 1 \equiv \boxed{1} \pmod{5}$$

Practice 17

Find the remainder when $9 \times 99 \times 999 \times \cdots \times \underbrace{99\cdots 9}_{999}$ is divided by 1000.

(ref: 280 - AIME)

Note that $\underbrace{99\cdots 9}_{k} \equiv 999 \equiv -1 \pmod{1000}$ when $k \geq 3$. The original expression is congruent to

$$9 \times 99 \times \underbrace{(-1)(-1)\cdots(-1)}_{997} \equiv 891 \times (-1) \equiv \boxed{109} \pmod{1000}$$

Practice 18

An integer N is selected at random in the range $1 \leq N \leq 2020$. What is the probability that the remainder when N^{16} is divided by 5 is 1?

(ref: 3743 - AMC10)

When $N \equiv 0 \pmod{5}$, then $N^{16} \equiv 0 \pmod{5}$.

When $N \equiv \pm 1 \pmod{5}$, then $N^{16} \equiv 1 \pmod{5}$.

When $N \equiv \pm 2 \pmod{5}$, then $N^{16} \equiv 2^{16} \equiv 4^8 \equiv (-1)^8 \equiv 1 \pmod{5}$.

We note that 2020 is a multiple of 5. Therefore, the answer is $4/5$.

Practice 19

Determine the units digit of the sum $0! + 1! + 2! + \cdots + n! + \cdots + 20!$?

(ref: 2654 - BCML)

Where $n \geq 5$, $n!$ must be a multiple of 10 whose units' digit will be 0. Therefore, the original question is equivalent to finding the unit digit of

$$0! + 1! + 2! + 3! + 4! \equiv 1 + 1 + 2 + 6 + 24 \equiv 34 \equiv \boxed{4} \pmod{10}$$

Practice 20

What are the last two digits in the sum of the factorials of the first 100 positive integers?

(ref: 1117)

When $k \geq 10$, $k!$ must be a multiple of 100 because its prime factorization contains two 5s and more than two 2s. Therefore, the desired result equals the last two digits of

$$1! + 2! + \cdots + 9! \pmod{100}$$

Let's compute these terms separately:

$$\begin{aligned}
1! &\equiv 1 &\pmod{100} \\
2! &\equiv 2 &\pmod{100} \\
3! &\equiv 6 &\pmod{100} \\
4! &\equiv 24 &\pmod{100} \\
5! &\equiv 20 &\pmod{100} \\
6! &\equiv 20 &\pmod{100} \\
7! &\equiv 40 &\pmod{100} \\
8! &\equiv 20 &\pmod{100} \\
9! &\equiv 80 &\pmod{100}
\end{aligned}$$

Adding the numbers on the right side leads to the result $\boxed{13}$.

Practice 21

Find the largest positive integer n such that $(3^{1024} - 1)$ is divisible by 2^n.

(ref: 2811)

Note that

$$3^{1024} - 1 = (3^{512} + 1)(3^{256} + 1)(3^{128} + 1) \cdots (3^2 + 1)(3 + 1)(3 - 1)$$

All the 11 factors are even. Among them

- $(3 - 1)$ is clearly not divisible by 4.
- $(3 + 1)$ is a multiple of 4.
- We claim none of the remaining terms is a multiple of 4.

The last conclusion holds because when k is a positive integer,

$$3^{2k} + 1 \equiv (-1)^{2k} + 1 \equiv 2 \pmod{4}$$

Therefore, $n = 1 + 2 + 9 = \boxed{12}$.

Practice 22

Let n be any positive integer, show that

$$(5n + 1)(5n + 2)(5n + 3)(5n + 4) \equiv -1 \pmod{25}$$

(ref: 4160)

$$
\begin{aligned}
& (5n+1)(5n+2)(5n+3)(5n+4) \\
=\ & ((5n+1)(5n+4))((5n+2)(5n+3)) \\
=\ & (25n^2+5n+4)(25n^2+5n+6) \\
=\ & (25n^2+5n)^2 + 10 \times (25n^2+5n) + 24 \\
\equiv\ & 24 \\
\equiv\ & -1 \pmod{25}
\end{aligned}
$$

Practice 23

Let N be the product of four consecutive odd numbers. Show that $N \equiv 1 \pmod 8$.

(ref: 4247)

Let these four consecutive numbers be $(2n-3)$, $(2n-1)$, $(2n+1)$, and $(2n+3)$ where integer $n \geq 2$. Then

$$
\begin{aligned}
& (2n-3)(2n-1)(2n+1)(2n+3) \\
=\ & (4n^2-9)(4n^2-1) \\
=\ & 16n^4 - 40n^2 + 9 \\
\equiv\ & 1 \pmod 8
\end{aligned}
$$

Practice 24

Let the product of all odd positive integer not greater than 2019 be 2019!!. Find the last three digits of 2019!!.

(ref: 4248)

Clearly, 2019!! is an odd number and multiple of 125. Therefore 2019!! can only end with 125, 375, 625, or 875. Their residues modulo 8 are 5, 7, 1, and 3, respectively.

By the conclusion of practice 23, we know the product of any four consecutive odd number is congruent to 1 modulo 8. Because 2019!! is a product of 1010 consecutive odd number and $1010 \equiv 2 \pmod 8$.

Therefore
$$2019!! \equiv 1 \times 3 \equiv 3 \pmod{8}$$

Hence the final answer is $\boxed{875}$.

Practice 25

Let n^2 be a square. Show that $n^2 \equiv 0, 1 \pmod{3}$.

(ref: 4206)

If $n \equiv 0 \pmod{3}$, then $n^2 \equiv 0 \pmod{3}$.

If $n \equiv \pm 1 \pmod{3}$, then $n^2 \equiv 1 \pmod{3}$.

Practice 26

Let k be a positive integer, show that $(4k + 3)$ cannot be a square number.

(ref: 2360)

A square can only be congruent to 0 or 1 modulo 4. Therefore, the claim holds.

Practice 27

Let n^2 be a square number. Show that $n^2 \equiv 0, \pm 1 \pmod{5}$.

(ref: 4156)

If $n \equiv 0 \pmod{5}$, then $n^2 \equiv 0 \pmod{5}$.

If $n \equiv \pm 1 \pmod{5}$, then $n^2 \equiv 1 \pmod{5}$.

If $n \equiv \pm 2 \pmod{5}$, then $n^2 \equiv 4 \equiv -1 \pmod{5}$.

Therefore, we conclude the claim holds.

Practice 28

Show that if n^2 is a square number, then $n^2 \equiv 0, 1, 4, 9 \pmod{16}$.

(ref: 4144)

If $n \equiv 0 \pmod{16}$, then $n^2 \equiv 0 \pmod{16}$.

If $n \equiv \pm 1 \pmod{16}$, then $n^2 \equiv 1 \pmod{16}$.

If $n \equiv \pm 2 \pmod{16}$, then $n^2 \equiv 4 \pmod{16}$.

If $n \equiv \pm 3 \pmod{16}$, then $n^2 \equiv 9 \pmod{16}$.

If $n \equiv \pm 4 \pmod{16}$, then $n^2 \equiv 16 \equiv 0 \pmod{16}$.

If $n \equiv \pm 5 \pmod{16}$, then $n^2 \equiv 25 \equiv 9 \pmod{16}$.

If $n \equiv \pm 7 \pmod{16}$, then $n^2 \equiv 49 \equiv 1 \pmod{16}$.

If $n \equiv 8 \pmod{16}$, then $n^2 \equiv 64 \equiv 0 \pmod{16}$.

Practice 29

Show that the difference of two squares of odd numbers must be a multiple of 8.

(ref: 2250)

Let n be an odd number, then $n \equiv \pm 1, \pm 3 \pmod{8}$ which leads to $n^2 \equiv 1 \pmod{8}$. Now it is obvious that the difference between squares of two odd numbers must be a multiple of 8.

Practice 30

How many terms in this sequence are squares?

$$1, 11, 111, 1111, \cdots$$

(ref: 4147)

The answer is $\boxed{1}$.

All these terms are odd. In order for any one to be a square, it must be congruent to 1 modulo 4. Only the first term 1 satisfies this condition. And clearly, 1 is a square. Hence, the answer is 1.

Practice 31

How many terms in this sequence are squares?

$$4, 44, 444, 4444, \cdots$$

(ref: 4148)

The answer is $\boxed{1}$.

All terms are even. Therefore, in order for any one to be a square, it must be congruent to 0 or 4 modulo 16. Only the first term satisfies this requirement. And indeed 4 is a square. (Note that the remainder when being divided by 16 is determined by its last three digits. So we only need to check the first three terms.)

This problem can also be reasoned using the conclusion of practice 30. Because 4 is a square number, therefore dividing every term of this sequence by 4 will not change the fact whether or not that term is a square. By practice 30, we know there is only one term in

$$1, 11, 111, 1111, \cdots$$

is square. So there is only one square in the original sequence as well.

Practice 32

How many numbers in this series are squares?

$$1, 14, 144, 1444, 14444, \cdots$$

(ref: 2361)

Among the first three terms, 1 and 144 are obviously square, but 14 is not.

We claim that none of the rest terms will be a perfect square. This can be verified by using the MOD 16 rule (see practice 28). All of the remaining terms are congruent to 12 modulo 16. Therefore, they cannot be squares. (Note that the remainder when being divided by 16 is determined by the last three digits.)

In conclusion, the answer to this question is $\boxed{2}$.

Practice 33

Find the number of integer pairs (x, y) such that $x^2 + y^2 = 2019$.

(ref: 4140)

The answer is $\boxed{0}$.

This is because $2019 \equiv 3 \pmod{4}$. But there exists no integer solution to the relation $x^2 + y^2 \equiv 3 \pmod{4}$.

Practice 34

Solve this equation in integers: $x_1^4 + x_2^4 + \cdots + x_{14}^4 = 9999$.

(ref: 4214)

First let's show that for any integer n, $n^4 \equiv 0, 1 \pmod{16}$.

- If $n \equiv 0, \pm 2, \pm 4, \pm 6, 8 \pmod{16}$, we have $n^4 \equiv 0 \pmod{16}$.

- If $n \equiv \pm 1, \pm 3, \pm 5, \pm 7 \pmod{16}$, we have $n^4 \equiv 1 \pmod{16}$.

Therefore, the value of $(x_1^4 + x_2^4 + \cdots + x_{14}^4) \pmod{16}$ can be 0, 1, $\cdots$, 14, but will never be 15. However $9999 \equiv 15 \pmod{16}$. This means that the given equation is insolvable.

Practice 35

What is the smallest positive integer greater than 5 which leaves a remainder of 5 when divided by each of 6, 7, 8, and 9?

(ref: 2645 - BCML)

It will be the least common multiple of 6, 7, 8, and 9 plus 5, which is 509.

Practice 36

A box contains gold coins. If the coins are equally divided among six people, four coins are left over. If the coins are equally divided among five people, three coins are left over. If the box holds the smallest number of coins that meets these two conditions, how many coins are left when equally divided among seven people?

(ref: 991 - AMC8)

We note that the remainder is always two less than the divisor when the coins are distributed among six or five people. Therefore, the number of coin must be two less than the least common multiple of 6 and 5, which is 28. It follows the answer is $\boxed{0}$.

Practice 37

Three runners start running simultaneously from the same point on a 500-meter circular track. They each run clockwise around the course maintaining constant speeds of 4.4, 4.8, and 5.0 meters per second. The runners stop once they are all together again somewhere on the circular course. How many seconds do the runners run?

(ref: 1395 - AMC10)

While MOD can only be applied to integer, but conceptually this problem is equivalent to solving

$$4.4t \equiv 4.8t \equiv 5.0t \pmod{500}$$

Subtracting $4.4t$ leads to

$$0 \equiv 0.4t \equiv 0.6t \pmod{500}$$

This means that we need to find a t such that $0.4t$ and $0.6t$ are both multiples of 500. The first few ts which satisfy this condition are 1250, 2500, $\cdots$. We find $\boxed{2500}$ is a solution.

Practice 38

Show that there is at least one Friday 13th in any year, including any leap year.

(ref: 4252)

This is equivalent to showing that every year, leap year included, has at least one Sunday the 1st each month. Let's exam the days when the 1st day of each month falls.

	Days of Year		MOD 7	
Month	Regular	Leap	Regular	Leap
January	1	1	1	1
February	32	32	4	4
March	60	61	4	5
April	91	92	0	1
May	121	122	2	3
June	152	153	5	6
July	182	183	0	1
August	213	214	3	4
September	244	245	6	0
October	274	275	1	2
November	305	306	4	5
December	335	336	6	0

Regardless of whether it is a leap year or not, the 1st day of each month includes all the residue classes of modulo 7. This means at least one of them will be on Sunday. Hence, the claim holds.

Practice 39

The number 2^{29} is a nine-digit number whose digits are all distinct. Which digit of 0 to 9 does not appear?

(ref: 311)

By the MOD by 9 technique, the sum of these digits must be congruent to 2^{29} (mod 9) which is (-4) (see below). Hence, the missing digit is $\boxed{4}$.

$$2^{29} \equiv (2^3)^9 \times 2^2 \equiv (-1)^9 \times 4 \equiv -4 \pmod 9$$

Practice 40

Select nine different digits from 0 to 9 to form a two-digit number, a three-digit number and a four-digit number. The sum of these three numbers is 2017. Which digit is not selected?

(ref: 4221)

Let the three numbers be $\overline{ab}$, $\overline{cde}$, and $\overline{fghi}$. Then
$$a + b + c + \cdots + i \equiv \overline{ab} + \overline{cde} + \overline{fghi} \equiv 2017 \equiv 1 \pmod{9}$$

Therefore, if there exists a solution, the missing digit must be 8. In fact, there exists a solution
$$43 + 269 + 1705 = 2017$$

Hence, the answer is $\boxed{8}$.

Practice 41

Find the multiplicative order of 3 modulo 17.

(ref: 4178)

It is sufficient to check divisors of $\varphi(17) = 16$.
$$\begin{aligned}
3^2 &\equiv 9 &\pmod{17} \\
3^4 &\equiv -4 &\pmod{17} \\
3^8 &\equiv -1 &\pmod{17} \\
3^{16} &\equiv 1 &\pmod{17}
\end{aligned}$$

Therefore the answer is $\boxed{16}$.

Practice 42

Find the multiplicative order of 5 modulo 19.

(ref: 4179)

It is sufficient to check divisors of $\varphi(19) = 18$.

$$5^2 \equiv 6 \pmod{19}$$
$$5^3 \equiv -8 \pmod{19}$$
$$5^6 \equiv 7 \pmod{19}$$
$$5^9 \equiv 1 \pmod{19}$$

Therefore the answer is $\boxed{9}$.

Practice 43

Find the multiplicative order of 2 modulo 125.

(ref: 4191)

By Euler's theorem, we have $2^{100} \equiv 1 \pmod{125}$. So it is sufficient to check all the divisors of 100: 2, 4, 5, 10, 20, 25, and 50.

$$2^2 \equiv 4 \pmod{125}$$
$$2^4 \equiv 16 \pmod{125}$$
$$2^5 \equiv 32 \pmod{125}$$
$$2^{10} \equiv 24 \pmod{125}$$
$$2^{20} \equiv 76 \pmod{125}$$
$$2^{25} \equiv 57 \pmod{125}$$
$$2^{50} \equiv -1 \pmod{125}$$

Therefore, we conclude the answer is $\boxed{100}$.

Practice 44

Let n be an odd integer greater than 1, then n is the multiplicative order of 2 modulo $(2^n - 1)$.

(ref: 4182)

It plainly holds that $2^n \equiv 1 \pmod{2^n - 1}$. Therefore, it is sufficient to show that for any positive integer k less than n, $2^k \not\equiv 1 \pmod{2^n - 1}$ which is equivalent to $(2^n - 1) \nmid (2^k - 1)$. This obviously holds because $(2^n - 1) > (2^k - 1) > 0$.

Practice 45

Let p be an odd prime, and integer a has multiplicative order of $2k$ modulo p, then $a^k \equiv -1 \pmod{p}$.

(ref: 4181)

Given $a^{2k} \equiv 1 \pmod{p}$, we have $p \mid (a^k+1)(a^k-1)$. Because p is an odd prime, it must hold that either $p \mid (a^k+1)$ or $p \mid (a^k-1)$ which means $a^k \equiv \pm 1 \pmod{p}$. However $p^k \equiv 1 \pmod{p}$ cannot hold because $2k$ is the multiplicative order. Therefore $a^k \equiv -1 \pmod{p}$.

Practice 46

Show that if integer a has multiplicative order of hk modulo n, then a^h has order of k modulo n.

(ref: 4180)

By the given condition, we have $a^{hk} \equiv 1 \pmod{n}$, but $a^m \not\equiv 1 \pmod{n}$ for any integer $0 < m < hk$.

Because $\left(a^h\right)^k \equiv a^{hk} \equiv 1 \pmod{n}$, we find the order of a^h modulo n is at most k. If there exists a positive integer r less than k such that $a^r \equiv 1 \pmod{n}$, we shall have $a^{rk} \equiv 1 \pmod{n}$ and $rk < hk$. This is impossible. Therefore the given claim must hold.

Practice 47

Let $n > 4$ be a composite number. Show that $(n-1)! \equiv 0 \pmod{n}$.

(ref: 4196)

Let p be the smallest prime divisor of n.

If $n \neq p^2$, then both p and n/p are less than n and distinct. Thus $(n-1)!$ is divisible by $p(n/p) = n$.

If $n = p^2$, then $p > 2$ because $n > 4$. It follows that both p and $2p$ are less than n. Therefore $(n-1)!$ is divisible by $p \cdot 2p = 2p^2 = 2n$, hence by n.

Practice 48

Let n be an integer greater than 1. If none of $1!, 2!, \cdots, n!$ has the same remainder when being divided by n, show that n is a prime.

(ref: 4202)

First, $n = 2, 3$ are both prime. When $n = 4$, we have $2! \equiv 3!$ (mod 4). By practice 47, we know for any composite $n > 4$, we have $(n-1)! \equiv 0 \pmod{n}$. This means $(n-1)! \equiv n! \equiv 0 \pmod{n}$.

Therefore we conclude the claim holds.

Practice 49

Show that for any positive integer k, it always holds that $10^k \equiv 4 \pmod{6}$.

(ref: 4224)

This conclusion can be proved in many different ways. Here is a solution using modular equations. Let $x = 10^k$. Then

$$\begin{cases} x = 10^k \equiv 0 \pmod{2} \\ x = 10^k \equiv 1 \pmod{3} \end{cases} \implies \begin{array}{l} x \equiv -2 \pmod{2} \\ x \equiv -2 \pmod{3} \end{array}$$

This means the solution is 2 less than the least common multiple of

2 and 3 which is 6. Or
$$10^k \equiv -2 \equiv 4 \pmod{6}$$

Practice 50

Show that for any integer x, the number $\left(\frac{x^5}{5} + \frac{x^3}{3} + \frac{7x}{15}\right)$ is an integer.

(ref: 4262)

This problem is equivalent to showing that $(3x^5 + 5x^3 + 7x)$ is always a multiple of 15. In order to show this, it is sufficient to show this expression is a multiple of both 3 and 5 for every x.

By Fermat little theorem, we have $x^3 \equiv x \pmod{3}$ and $x^5 \equiv x \pmod{5}$. Therefore

$$3x^5 + 5x^3 + 7x \equiv 0 + 2x^3 + x \equiv 2x + x \equiv 3x \equiv 0 \pmod{3}$$
$$3x^5 + 5x^3 + 7x \equiv 3x + 0 + 2x \equiv 5x \equiv 0 \pmod{5}$$

Therefore, we conclude the claim holds.

Practice 51

Show that $(2^{1194} + 1)$ is a multiple of 65.

(ref: 4167)

Because 65 can be prime factorized into 5×13, it is sufficient to show the given number is both a multiple of 5 and a multiple of 13.

By Fermat's little theorem, we have $2^4 \equiv 1 \pmod{5}$ and $2^{12} \equiv 1 \pmod{13}$. Therefore

$$2^{1194} - 1 \equiv \left(2^4\right)^{298} \times 2^2 + 1 \equiv 0 \pmod{5}$$

and

$$2^{1194} - 1 \equiv \left(2^{12}\right)^{99} \times 2^6 + 1 \equiv 0 \pmod{13}$$

Therefore, the claim holds.

Practice 52

Let $k = 2008^2 + 2^{2008}$. What is the units digit of $k^2 + 2^k$?

(ref: 1608 - AMC10)

Firstly, 2008^2 must end with 4. Meanwhile the units digit of 2^k repeats every 4 numbers: 2, 4, 8, 6, 2, $\cdots$. This means that the end digit of 2^{2008} must end with 6. Therefore, k ends with 0 which implies k^2 ends with 0.

To determine the last digit of 2^k, it is sufficient to compute k (mod 4) because we can utilize the repeating pattern observed above. It is easy to see that $k \equiv 0 \pmod{4}$. Therefore, 2^k will end with 6.

Hence $k^2 + 2^k$ ends with $\boxed{6}$.

Practice 53

Compute 9^{50} (mod 1000).

(ref: 4216)

This problem can be solved with the binomial expansion.

$$9^{50} = (10-1)^{50} = \cdots - C_{50}^3 \times 10^3 + C_{50}^2 \times 10^2 - C_{50}^1 \times 10 + C_{50}^0$$

It is clear that all terms except the last three are multiple of 1000. Therefore their sum must be congruent to 0 modulo 1000. Therefore

$$9^{50} \equiv C_{50}^2 \times 10^2 - C_{50}^1 + C_{50}^0 \equiv \frac{50 \times 49}{2} \times 100 - 500 + 1 \equiv \boxed{1} \pmod{1000}$$

Practice 54

Find the last three digits of $9 + 9^2 + 9^3 + \cdots + 9^{2000}$.

(ref: 4217)

By practice 53, we have $9^{50} \equiv 1 \pmod{1000}$. Therefore, we have

$$9 \equiv 9^{51} \equiv 9^{101} \equiv \cdots, 9^2 \equiv 9^{52} \equiv 9^{102} \equiv \cdots$$

This means

$$9 + \cdots + 9^{50} \equiv 9^{51} + \cdots + 9^{100} \equiv \cdots + 9^{1951} + \cdots + 9^{2000} \pmod{1000}$$

Noting $9^{51} \equiv 9 \pmod{1000}$.

$$9 + 9^2 + \cdots + 9^{50} = \frac{9^{51} - 9}{9 - 1} \equiv 0 \times 8^{-1} \equiv 0 \pmod{1000}$$

Therefore the desired answer is $40 \times 0 \equiv \boxed{0} \pmod{1000}$

Practice 55

Find the remainder when $10^{10} + 10^{100} + 10^{1000} + \cdots + 10^{\overbrace{10 \cdots 0}^{2018}}$ is divided by 7.

(ref: 4225)

Because $\varphi(7) = 6$ and $(10, 7) = 1$, we find $10^6 \equiv 1 \pmod 7$ by Euler's theorem.

Meanwhile by the conclusion of practice 49, we have $10^k \equiv 4 \pmod 6$ hold for any integer k. Therefore

$$10^{10^m} \equiv 10^{6n} \times 10^4 \equiv 10^4 \pmod 7$$

holds for any positive integer m and corresponding positive integer

n. It follows that

$$\begin{aligned}
& 10^{10} + 10^{100} + 10^{1000} + \cdots + 10^{\overbrace{10\cdots 0}^{2018}} \\
\equiv\ & 10^4 + 10^4 + \cdots + 10^4 \\
\equiv\ & 10^4 \times 2018 \\
\equiv\ & 3^4 \times 2 \\
\equiv\ & 4 \times 2 \\
\equiv\ & \boxed{1} \pmod 7
\end{aligned}$$

Practice 56

Given 30! ends with some zeros, what is the digit that immediately precedes these zeros?

(ref: 4173)

Let's first factorize:

$$30! = 2^{26} \times 3^{14} \times 5^7 \times 7^4 \times 11^2 \times 13^2 \times 17 \times 19 \times 23 \times 29$$

Hence, 30! will end with 7 zeros. Removing these zeros will get

$$N = \frac{30!}{10^7} = 2^{21} \times 3^{14} \times 7^4 \times 11^2 \times 13^2 \times 17 \times 19 \times 23 \times 29$$

The desired result is the last digit of N:

$$\begin{aligned}
N &\equiv 2^{21} \times 3^{14} \times 7^4 \times 1^2 \times 3^2 \times 7 \times 9 \times 3 \\
&\equiv 2^{21} \times 3^{19} \times 7^5 \\
&\equiv \boxed{8} \pmod{10}
\end{aligned}$$

Practice 57

The two-digit integers from 19 to 92 are written consecutively to form the large integer

$$N = 192021\cdots 909192$$

Suppose that the 3^k is the highest power of 3 that is a factor of N. What is k.

(ref: 4174 - AMC12)

The answer is $\boxed{1}$. We are going to show that N is a multiple of 3, but not a multiple of 9. It can be computed that the sum of N's digits is 705. Then it is clear now that this conclusion holds because $S(N) \equiv 3 \pmod 9$.

Practice 58

Let S be the sum of squares of 10 consecutive positive integers. Show S cannot be a square.

(ref: 4250)

Let the 10 consecutive numbers be $n, (n+1), \cdots, (n+9)$. Then, its sum equals

$$S = n^2+(n+1)^2+\cdots+(n+9)^2 = 10n^2+90n+285 = 5\times(2n^2+18n+57)$$

Therefore, it is sufficient to show that $(2n^2 + 18n + 57)$ is not a multiple of 5. If so, S is a multiple of 5 but not 5^2, hence cannot be a square.

- If $n \equiv 0 \pmod 5$, then $2n^2 + 18n + 57 \equiv 2 \pmod 5$.
- If $n \equiv 1 \pmod 5$, then $2n^2 + 18n + 57 \equiv 2 \pmod 5$.
- If $n \equiv 2 \pmod 5$, then $2n^2 + 18n + 57 \equiv 1 \pmod 5$.

- If $n \equiv -1 \pmod 5$, then $2n^2 + 18n + 57 \equiv 1 \pmod 5$.
- If $n \equiv -2 \pmod 5$, then $2n^2 + 18n + 57 \equiv 4 \pmod 5$.

Therefore, we conclude the claim holds.

Practice 59

What is the remainder when $(8888^{2222} + 7777^{3333})$ is divided by 37?

(ref: 167)

Because

- $8888 \equiv 8 \pmod{37}$
- $7777 \equiv 7 \pmod{37}$
- $8^2 \equiv -10 \pmod{37}$
- $7^3 \equiv 10 \pmod{37}$

$\therefore 8888^{2222} + 7777^{3333} \equiv 8^{2222} + 7^{3333} \equiv (-10)^{1111} + 10^{1111} \equiv 0 \pmod{37}$

This means the answer is $\boxed{0}$.

Practice 60

Compute $3^{2018} \bmod 17$.

(ref: 4161)

By Fermat's little theorem, we have $3^{16} \equiv 1 \pmod{17}$. Therefore

$$3^{2018} \equiv \left(3^{16}\right)^{126} \times 3^2 \equiv \boxed{9} \pmod{17}$$

Practice 61

Compute 50^{250} (mod 83).

(ref: 2740)

By Fermat's little theorem, we have $50^{82} \equiv 1$ (mod 83) because 83 is prime and $(83, 50) = 1$.

$50^{250} = 50^{246} \cdot 50^4 = \left(50^{82}\right)^3 \times 2500^2 \equiv 1^3 \times 10^2 = 100 \equiv \boxed{17}$ (mod 83).

Practice 62

Compute 20! (mod 23).

(ref: 4169)

By Wilson's theorem, we have $22! \equiv -1$ (mod 23). Therefore

$1 \equiv 22! \equiv 22 \times 21 \times 20! \equiv (-1) \times (-2) \times 20! \equiv 2 \times 20!$ (mod 23)

By noting 12 is the inverse of 2 modulo 23, we find

$$20! \equiv 12 \times (-1) \equiv \boxed{11} \pmod{23}$$

Practice 63

Let $N = 7 \times 8 \times 9 \times 15 \times 16 \times 17 \times 23 \times 24 \times 25 \times 43$. Compute N (mod 11).

(ref: 4218)

We have

$$\begin{aligned}
&7 \times 8 \times 9 \times 15 \times 16 \times 17 \times 23 \times 24 \times 25 \times 43 \\
\equiv\ &7 \times 8 \times 9 \times 4 \times 5 \times 6 \times 1 \times 2 \times 3 \times 10 \\
\equiv\ &10! \\
\equiv\ &-1 \pmod{11}
\end{aligned}$$

The last step is the result of applying the Wilson's theorem. Therefore the answer is $\boxed{10}$.

Practice 64

Let p is an odd prime, compute $1^{p-1}+2^{p-1}+3^{p-1}+\cdots+(p-1)^{p-1}$ (mod p).

(ref: 4228)

By Fermat's little theorem, we have

$$\begin{aligned}
\therefore\ & 1^{p-1} + 2^{p-1} + 3^{p-1} + \cdots + (p-1)^{p-1} \\
\equiv\ & 1 + 1 + 1 + \cdots + 1 \\
\equiv\ & p - 1 \\
\equiv\ & \boxed{-1} \quad (\text{mod } p)
\end{aligned}$$

Practice 65

Let p is an odd prime, compute $1^p + 2^p + 3^p + \cdots + (p-1)^p$ (mod p).

(ref: 4229)

By Fermat's little theorem, we have

$$\begin{aligned}
\therefore\ & 1^p + 2^p + 3^p + \cdots + (p-1)^p \\
\equiv\ & 1 + 2 + 3 + \cdots + (p-1) \\
\equiv\ & p(p-1)/2 \\
\equiv\ & \boxed{0} \quad (\text{mod } p)
\end{aligned}$$

Practice 66

Find the smallest positive integer n so that $107n$ has the same last two digits as n.

(ref: 2805 - Harvard-MIT)

This is equivalent to solving the following equation

$$n \equiv 107n \pmod{100}$$

or

$$n \equiv 7n \pmod{100} \implies 6n \equiv 0 \pmod{100}$$
$$\therefore \quad 6n = 100k \implies n = 50 \cdot \frac{k}{3}$$

where k is an integer. Clearly, the smallest such positive integer is $\boxed{50}$.

Practice 67

How many positive integers N, less than 2017, satisfy

$$N^{2016^{2016}} \equiv 1 \pmod{2017}$$

(ref: 4227)

Because 2017 is a prime number. All positive integers less than 2017 are co-prime to it. Meanwhile, $\varphi(2017) = 2016$. Therefore by Euler's theorem, we have

$$N^{2016} \equiv 1 \pmod{2017} \implies (N^{2016})^{2016} \equiv 1 \pmod{2017}$$

This means all such Ns satisfy the requirement. Hence, the answer is $\boxed{2016}$.

Practice 68

Solve $x^{12} \equiv 3 \pmod{11}$.

(ref: 4164)

Because 11 is a prime, any solution must be co-prime to 11. Then by Fermat's little theorem, we have $x^{10} \equiv 1 \pmod{11}$. Then it follows that $x^2 \equiv 3 \pmod{11}$. We test $x = 0, \pm 1, \pm 2, \pm 3, \pm 4, \pm 5$ and find $x \equiv \pm 5$ are the solutions.

Practice 69

Solve this modular equation:
$$f(x) = 4x^2 + 27x - 9 \equiv 0 \pmod{15}$$

(ref: 4240)

For this equation to be solvable, both $f(x) \equiv 0 \pmod 3$ and $f(x) \equiv 0 \pmod 5$ need to be solvable.

Considering $\pmod 3$ first.
$$f(x) = 4x^2 + 27x - 9 \equiv x^2 \equiv 0 \pmod 3 \implies x \equiv 0 \pmod 3$$

This means that $f(x) \equiv 0 \pmod 3$ has only one solution $x \equiv 0 \pmod 3$. Now setting $x = 0$, an element in this residue class, to $f(x) \equiv 0 \pmod 5$ yields:
$$f(0) = -9 \equiv 1 \not\equiv 0 \pmod 5$$

This implies that $f(x) \equiv 0 \pmod 3$ and $f(x) \equiv 0 \pmod 5$ cannot hold simultaneously. Hence, we conclude that $f(x) \equiv 0 \pmod{15}$ is insolvable.

Practice 70

Compute $3^{2017} \pmod{1000}$.

(ref: 4241)

This is equivalent to finding the least non-negative integer x satisfying
$$\begin{cases} x \equiv 3^{2017} \equiv 3 & \pmod 8 \\ x \equiv 3^{2017} \equiv 38 & \pmod{125} \end{cases}$$

Applying CRT:
$$\begin{aligned} 125^{-1} &\equiv 125^{\varphi(8)-1} \equiv 5 & \pmod 8 \\ 8^{-1} &\equiv 8^{\varphi(125)-1} \equiv 47 & \pmod{125} \end{aligned}$$

$$\therefore x \equiv 3 \times 125 \times 5 + 38 \times 8 \times 47 \equiv \boxed{163} \pmod{1000}$$

Practice 71

Solve the system of congruence

$$\begin{cases} x \equiv 1 \pmod{3} \\ x \equiv 2 \pmod{5} \\ x \equiv 3 \pmod{7} \end{cases}$$

(ref: 4197)

Because 3, 5, and 7 are pair-wise co-prime, we can apply the CRT.

$$3 \times 5 \times 7 = 105$$
$$(5 \times 7)^{-1} \equiv 35^{-1} \equiv 2 \pmod{3}$$
$$(7 \times 3)^{-1} \equiv 21^{-1} \equiv 1 \pmod{5}$$
$$(3 \times 5)^{-1} \equiv 15^{-1} \equiv 1 \pmod{7}$$

Therefore the solution is

$$x \equiv 1 \times 35 \times 2 + 2 \times 21 \times 1 + 3 \times 15 \times 1 \equiv 52 \pmod{105}$$

Practice 72

Show that if there exist integer x, y, and z such that $3^x + 4^y = 5^z$, then both x and z must be even.

(ref: 4185)

Taking MOD 4 on both sides gives $(-1)^x + 0 \equiv 1^z \pmod{4}$. It can only hold if x is even.

Next, taking MOD 3 on both sides yields $0 + 1^y \equiv (-1)^z \pmod{3}$. Hence, z must be even.

Practice 73

Solve the congruent system: $4x \equiv 2 \pmod 6$ and $3x \equiv 5 \pmod 8$.

(ref: 4199)

Let's start by simplifying the first equation to $2x \equiv 1 \pmod 3$. While 3 and 8 are now co-prime, we need to adjust the coefficient to 1 in order to apply the CRT. Note $2^{-1} \equiv 2 \pmod 3$ and $3^{-1} \equiv 3 \pmod 8$, we have

$$\begin{cases} x \equiv 2 & \pmod 3 \\ x \equiv -1 & \pmod 8 \end{cases}$$

Applying CRT gives

$$x \equiv 2 \times 8 \times 2 + (-1) \times 3 \times 3 \equiv 23 \pmod{24}$$

Additionally, we can rewrite the first relation as $x \equiv 2 \equiv -1 \pmod 3$. Then, the solution must be one less the least common multiple of 3 and 8 which leads to the same result.

Practice 74

Show that $2x^2 - 5y^2 = 7$ has no integer solution.

(ref: 125)

It is clear that y is an odd number. Hence $y^2 \equiv 1 \pmod 8$ and $y^2 \equiv 1 \pmod 4$. Rewriting the given equation as

$$2x^2 = 5y^2 + 7$$

If x is an even number, then $2x^2 \equiv 0 \pmod 8$. But $5y^2 + 7 \equiv 5 + 7 \equiv 4 \pmod 8$. It is a contradiction.

If x is an odd number, then $2x^2 \equiv 2 \pmod 4$. But $5y^2 + 7 \equiv 0$

(mod 4). It is a contradiction too.

Practice 75

Let m be the least positive integer divisible by 17 whose digits sum is 17. Find m.

(ref: 70 - AIME)

Let $s(m)$ be the digit sum of m. It must hold that

$$m \equiv s(m) \equiv 17 \equiv -1 \pmod 9$$

Because m is a multiple of 17, let $m = 17k$ which means

$$17k \equiv -1 \pmod 9$$

Obviously $k = 1$ is one solution, therefore its general solution is $k = 9n + 1$. Try $n = 0, 1, 2, 3, \cdots$ finding $m = \boxed{476}$ is the first positve solution that has $s(m) = 17$.

Practice 76

Let $f(n)$ denote the sum of the digits of n. Find $f(f(f(4444^{4444})))$.

(ref: 2216 - IMO)

This is a typical problem that can be solved by the MOD-by-9 method.

Because $4444^{4444} < 10000^{4444} = 10^{17776}$, we find that 4444^{4444} has at most 17776 digits, which means that $f(4444^{4444})$ can not be greater than $9 \times 17776 = 159984$. It follows that $f(f(4444^{4444}))$ can not be greater than $9 \times 5 = 45$. Similarly, $f(f(f(4444^{4444})))$ can not be greater than $3 + 9 = 12$.

Meanwhile, we have

$$f(f(f(4444^{4444}))) \equiv 4444^{4444} \pmod 9$$

and

$$4444^{4444} \equiv (-2)^{4444} = 2^{4440} \times 2^4 = 64^{740} \times 16 \equiv 1^7 \times 7 \equiv 7 \pmod 9$$

There is only one positive integer no greater than 12 which is congruent to 7 modulo 9. Hence, the answer is $\boxed{7}$.

Practice 77

What is the last digit of $17^{17^{17^{17}}}$?

(ref: 2540 - PUMaC)

We know that the last digit of 17^k repeats as 7, 9, 3, 1, 7, $\cdots$. Therefore it is sufficient to compute $17^{17^{17}} \pmod 4$ which is

$$17^{17^{17}} \equiv 1^{17^{17}} \equiv 1 \pmod 4$$

Therefore, the final answer is $\boxed{7}$.

Practice 78

Let $N = 4568^{7777}$, a be the sum of digits in N, b be the sum of digits in a, and c be the sum of digits in b. Find c.

(ref: 4261)

Because $N = 4568^{7777} < 10000^{7777} = 10^{31109}$, therefore N has at most 31109 digits. It follows that $a < 9 \times 31109 = 27998$. Then, we find $b < 2 + 5 \times 9 = 47$, and $c < 4 + 9 = 13$.

Meanwhile, we have

$$c \equiv 4568^{7777} \equiv 5^{7777} \equiv (5^3)^{2592} \times 5 \equiv (-1)^{2592} \times 5 \equiv 5 \pmod 9$$

Therefore, we find $c = \boxed{5}$.

Practice 79

Let $\mathbb{S}$ be a set containing all the integers created by digits 1, 2, $\cdots$, 7. Each digit can be used once and only once. Show that no element in $\mathbb{S}$ is a multiple of the other.

(ref: 4208)

If this claim is not true, let two numbers $a, b \in \mathbb{S}$ and $a = bc$ where c is a positive integer $c > 1$. Let $S(n)$ be the sum of digits of n, then
$$S(a) \equiv S(b) \equiv 1 + 2 + \cdots + 7 \equiv 1 \pmod{9}$$

Now we have
$$a = bc \implies 1 \equiv 1 \times c \pmod{9} \implies c \equiv 1 \pmod{9}$$

Because $c > 1$, then $c \geq 10$. However this cannot be held since both b and c are seven-digit numbers.

Practice 80

Let p be a prime and integer a is co-prime to p, show that
$$a^{p(p-1)} \equiv 1 \pmod{p^2}$$

(ref: 4170)

It is clear that a will be co-prime to p^2. Meanwhile, we have $\varphi(p^2) = p(p-1)$. Therefore, by Euler's theorem, we have $a^{p(p-1)} \equiv 1 \pmod{p^2}$.

Practice 81

Let p and q be two distinct primes, and integer a is co-prime to both p and q, show

$$a^{(p-1)(q-1)} \equiv 1 \pmod{pq}$$

(ref: 4171)

It is clear that a is co-prime to pq and $\varphi(pq) = (p-1)(q-1)$. Therefore by Euler's theorem, we have

$$a^{(p-1)(q-1)} \equiv 1 \pmod{pq}$$

Practice 82

Show that two positive integers m and n are co-prime if and only if $\varphi(mn) = \varphi(m)\varphi(n)$.

(ref: 4195)

Let p_i be the common prime divisors of m and n, if any. Let q_i be those prime factors which divide m, but not n. Let r_i be those prime factors which divide n, but m. Then we have

$$m = \prod p_i^{\alpha_i} \prod q_j^{\beta_j}$$

and

$$n = \prod p_i^{\gamma_i} \prod r_j^{\delta_j}$$

where α_i, β_i, γ_i and δ_i are all non-negative integers. Then we have

$$\varphi(mn) = mn \prod \left(1 - \frac{1}{p_i}\right) \prod \left(1 - \frac{1}{q_j}\right) \prod \left(1 - \frac{1}{r_k}\right)$$

and

$$\varphi(m)\varphi(n) = mn \prod \left(1 - \frac{1}{p_i}\right)^2 \prod \left(1 - \frac{1}{q_j}\right) \prod \left(1 - \frac{1}{r_k}\right)$$

Now, it is clear that these two will equal if and only if the term $\prod\left(1-\frac{1}{p_i}\right)$ vanishes which means that m and n do not share any common prime divisors, i.e. they are co-prime.

Practice 83

Let p be a prime and k be a positive integer less than p. Show that $C_p^k \equiv 0 \pmod{p}$.

(ref: 4232)

Because p is a prime and $k < p$, we must have $p \nmid k!$. Meanwhile, because $k > 0$, we have $(p - k) < p$ which means $p \nmid (p - k)!$.

However, $C_p^k = \frac{p!}{k!(p-k)!}$ is an integer. The denominator, $p!$, is obviously a multiple of p, but the numerator is not. Hence, the result must be a multiple of p.

Practice 84

Show that $\varphi(n) = n/4$ is impossible to hold.

(ref: 4220)

Suppose there exists an integer n such that $\varphi(n) = n/4$. Then n must be a multiple of 4 because $\varphi(n)$ is an integer. Let the prime factorization of n be

$$n = 2^m p_1^{k_1} p_2^{k_2} \cdots p_j^{k_j}$$

where $m \geq 2$. Now

$$\varphi(n) = 2^{m-1} p_1^{k_1-1} p_2^{k_2-1} \cdots p_j^{k_j-1}(p_1 - 1)(p_2 - 1) \cdots (p_j - 1)$$

Setting $\varphi(n) = n/4$ and canceling common terms will yield

$$p_1 p_2 \cdots p_j = 2(p_1 - 1)(p_2 - 1) \cdots (p_j - 1)$$

But this relation cannot hold because the left side is odd and the right side is even.

Practice 85

Let $P(x)$ be a polynomial with integer coefficients satisfying that both $P(0)$ and $P(1)$ are odd. Show that $P(x)$ has no integer zeros.

(ref: 2841)

If this is not true, then there exists an integer k such that $P(k) = 0$.

If k is even, then $P(k) \equiv P(0) \equiv 1 \pmod{2}$. And, if k is odd, then $P(x) \equiv P(1) \equiv 1 \pmod{2}$. Therefore, regardlessly, it always hold that
$$P(k) \equiv 1 \pmod{2}$$

This means $P(k) \neq 0$, or k is not a zero of $P(x)$.

Practice 86

Does there exist a polynomial $P(x)$ such that $P(1) = 2015$ and $P(2015) = 2016$?

(ref: 3971)

Firstly, we know that if k is odd, then for any integer m, we must have $km \equiv m \pmod{2}$.

Let $P(x) = a_n x^n + a_{n-1} x^{n-1} + \cdots + a_1 x + a_0$. Then

$$P(2015) \equiv a_n + a_{n-1} + \cdots + a_1 + a_0 = P(1) = 2015 \equiv 1 \pmod{2}$$

But 2016 is even, Therefore it is impossible for $P(2015) = 2016$.

Practice 87

Find 8 prime numbers, not necessarily distinct such that the sum of the squares of these numbers is 992 less than 4 times of the product of these numbers.

(ref: 2822)

We need to solve for prime numbers, p_i, $i = 1, 2, \cdots, 8$ such that

$$\sum_{i=1}^{8} p_i^2 + 992 = 4 \prod_{i=1}^{8} p_i$$

If any p_i is odd, then $p_i^2 \equiv 1 \pmod 8$. Therefore, if all the p_i are odd, the left side of the above equation is a multiple of 8, but the right side is not. This means that at least one of p_i must be even.

However, if at least one of p_i is even, the right side is divisible by 8. In this case, the left side cannot be a multiple of 8 unless all the p_i are even.

There is only one even prime which is 2. Setting all p_i as 2 does satisfy the above equation. Therefore, the only solution is

$$p_1 = p_2 = \cdots = p_8 = 2$$

Practice 88

Show that for any positive integer n, $\varphi(2^n - 1)$ is a multiple of n where $\varphi(n)$ is Euler's totient function.

(ref: 4183)

By practice 44, we have n is the multiplicative order of 2 modulo $(2^n - 1)$. Meanwhile, by Euler's theorem, we have $2^{\varphi(2^n-1)} \equiv 1 \pmod{2^n - 1}$. Therefore, we have $n \mid \varphi(2^n - 1)$.

Practice 89

Let p be an odd prime divisor of integer $(n^4 + 1)$. Show that $p \equiv 1 \pmod 8$.

(ref: 4184)

Given p divides $(n^4 + 1)$, we have $n^4 \equiv -1 \pmod p$ and $n^8 \equiv 1 \pmod p$.

It is clear that p and n are co-prime, therefore by Euler's theorem, we have $n^{\varphi(p)} \equiv n^{p-1} \equiv 1 \pmod p$. Therefore $8 \mid (p-1)$ or $p \equiv 1 \pmod 8$.

Practice 90

Let n be a positive odd integer. Show that at least one of the following numbers is a multiple of n.

$$2-1, 2^2-1, \cdots, 2^{n-1}-1$$

(ref: 4204)

The claim obviously hold when $n = 1$. When $n > 1$, let's consider the following n numbers:

$$1, 2, \cdots, 2^{n-1}$$

They are all powers of 2, but n is odd, therefore none of them is a multiple of n. It follows that at least two of them are congruent modulo n. Let these two numbers be a^i and 2^j, where $0 \le i < j \le n-1$. Now we have

$$2^i \equiv 2^j \pmod n \implies 2^i(2^{j-i} - 1) \equiv 0 \pmod n$$

Again, $2^i \not\equiv 0 \pmod n$. Therefore $2^{i-j} \equiv 0 \pmod n$ where $1 \le i - j \le n - 1$.

Practice 91

Show that from any given m integers, it is always possible to select one or more integers such that their sum is a multiple of m.

(ref: 4223)

Let these m integers be $a_1, a_2, \cdots, a_m$. Then consider the following m sums:
$$\begin{aligned} S_1 &= a_1 \\ S_2 &= a_1 + a_2 \\ S_3 &= a_1 + a_2 + a_3 \\ &\cdots \\ S_m &= a_1 + a_2 + a_3 + \cdots + a_m \end{aligned}$$

If any of them is a multiple of m, the conclusion already holds. Otherwise, by the pigeonhole principle, at least two of them must have the same residue of modulo m. Hence, their difference, which is also a sum of some a_i, is a multiple of m.

Practice 92

Let sequence $\{x_n\}$ satisfy the relation $x_{n+2} = x_{n+1} + 2x_n$ for $n \geq 1$ where $x_1 = 1$ and $x_2 = 3$.

Let sequence $\{y_n\}$ satisfy the relation $y_{n+2} = 2y_{n+1} + 3y_n$ for $n \geq 1$ where $y_1 = 7$ and $y_2 = 17$.

Show that these two sequences do not share any common term.

(ref: 4209)

We are going to show that both $\{x_n \pmod 8\}$ and $\{y_n \pmod 8\}$ are cyclic after certain terms.

For $\{x_n\}$, we have $x_2 \equiv 3 \pmod 8$ and $x_3 = 3 + 2 \times 1 \equiv 5 \pmod 5$.

For $n \geq 4$, we have
$$\begin{cases} x_{n+2} = x_{n+1} + 2x_n & \equiv 5 + 2 \times 3 \equiv 3 \pmod{8} \\ x_{n+3} = x_{n+2} + 2x_{n+1} & \equiv 3 + 2 \times 5 \equiv 5 \pmod{8} \end{cases}$$

Therefore, we conclude $\{x_n \pmod{8}\} = \{1, 3, 5, 3, 5, 3, 5, \cdots\}$.

Meanwhile, for $\{y_n\}$, we have $y_1 \equiv 7 \pmod{8}$ and $y_2 \equiv 1 \pmod{8}$. When $n \geq 3$, we have
$$\begin{cases} y_{n+2} = 2y_{n+1} + 3y_n & \equiv 2 \times 7 + 3 \times 1 \equiv 1 \pmod{8} \\ y_{n+3} = 2y_{n+2} + 3y_{n+1} & \equiv 2 \times 1 + 3 \times 7 \equiv 7 \pmod{8} \end{cases}$$

Therefore, we conclude $\{y_n \pmod{8}\} = \{7, 1, 7, 1, 7, 1, \cdots\}$.

Clearly, $\{x_n\}$ and $\{y_n\}$ do not share some initial terms. Therefore no terms will be equal in these two sequences.

Practice 93

Show that if n is an integer greater than 1, then $(2^n - 1)$ is not divisible by n.

(ref: 3624 - Putnam)

If this claim does not hold, let's assume there exists an integer $n > 1$ such that $n \mid (2^n - 1)$. Then n must be an odd number because $(2^n - 1)$ is odd.

Let p be the least prime divisor of n, then $n \mid (2^n - 1)$ implies $p \mid (2^n - 1)$, or equivalently, $2^n \equiv 1 \pmod{p}$. By Fermat's little theorem, we have $2^{p-1} \equiv 1 \pmod{p}$. Let $d = gcd(n, p-1)$ then $2^d \equiv 1 \pmod{p}$. By the definition of p, since $d \mid n$ and $d \leq p - 1 < p$, we get $d = 1$. Then $2 = 2^d \equiv 1 \pmod{p}$. This is a contradiction. Therefore, the previous assumption does not hold which means $(2^n - 1)$ is not divisible by n.

Practice 94

Let p be an odd prime divisor of number $(a^2 + 1)$ where a is an integer. Show that $p \equiv 1 \pmod 4$.

(ref: 3870)

Because $p \mid a^2 + 1$, therefore $a^2 \equiv -1 \pmod p$ which implies $a^4 \equiv 1 \pmod p$.

It follows that a and p are co-prime. Therefore, by Euler's theorem, we have $a^{\varphi(p)} \equiv 1 \pmod p$, or $a^{p-1} \equiv 1 \pmod p$. Thus, $4 \mid (p-1)$ or $p \equiv 1 \pmod 4$.

Practice 95

Let a and b be two positive integers such that both of them can be written as a sum of two squares. Show that their product can be written as a sum of two squares in two ways.

(ref: 4189)

Let $a = x^2 + y^2$ and $b = z^2 + t^2$ where x, y, z, and t are all integers. Then we have

$$\begin{aligned} ab &= (x^2 + y^2)(z^2 + t^2) \\ &= x^2z^2 + x^2t^2 + y^2z^2 + y^2t^2 \\ &= (x^2z^2 + 2xyzt + y^2t^2) + (x^2t^2 - 2xyzt + y^2z^2) \\ ab &= (xz + yt)^2 + (xt - yz)^2 \\ ab &= (xz - yt)^2 + (xt + yz)^2 \end{aligned}$$

Practice 96

Let $\{a_1, a_2, \cdots, a_{2n+1}\}$ be a set of integers such that after removing any element, the remaining ones can always be equally divided into two groups with equal sum. Show that all these a_i, $(1 \le i \le 2n+1)$ are equal.

(ref: 4255 - Putnam)

If the sum of the remaining $2n$ elements is always even, regardless of which element is removed, then all the elements must have the same parity. This means

$$a_1 \equiv a_2 \equiv \cdots \equiv a_{2n+1} \pmod{2}$$

Let $a_i^{(2)} = \frac{a_i}{2}$ or $\frac{a_i - 1}{2}$ depending on the parity of a_i where $1 \le i \le n$. Clearly, after removing any $a_i^{(2)}$, the remaining elements can also be equally divided into two groups with equal sum. This means all these $a_i^{(2)}$ have the same parity. This means that

$$a_1^{(2)} \equiv a_2^{(2)} \equiv \cdots \equiv a_{2n+1}^{(2)} \pmod{2}$$
$$\implies a_1 \equiv a_2 \equiv \cdots \equiv a_{2n+1} \pmod{2^2}$$

This process can be repeatedly applied and as a result the following congruence will hold for every positive integer k:

$$a_1 \equiv a_2 \equiv \cdots \equiv a_{2n+1} \pmod{2^k}$$

This cannot be true unless these a_i are all equal.

Practice 97

Let x and y be two integers and p be a prime. Show that

$$(x+y)^p \equiv x^p + y^p \pmod{p}$$

(ref: 4233)

By practice 83, we have $C_p^k \equiv 0 \pmod{p}$ for all the positive integer k less than p. Therefore, we have
$$\begin{aligned}(x+y)^p &= x^p + C_p^1 x^{p-1} y + \cdots + C_p^{p-1} x y^{p-1} + y^p \\ &\equiv x^p + 0 + \cdots + 0 + y^p \\ &\equiv x^p + y^p \pmod{p}\end{aligned}$$

Practice 98

Let n be a positive integer and k be an odd positive integer, show $k^{2^n} \equiv 1 \pmod{2^{n+2}}$.

(ref: 4243)

Apply mathematical induction on n.

When $n = 1$. Because k is odd, therefore $k \equiv \pm 1, \pm 3 \pmod{8} \implies k^2 \equiv 1 \pmod{8}$. The claim holds.

Assume the claim hold when $n = m \geq 1$, i.e. $k^{2^m} \equiv 1 \pmod{2^{m+2}}$. Let $k^{2^m} - 1 = 2^{m+2} \cdot t$ where t is an integer.

Then when $n = m + 1$, we have
$$\begin{aligned}k^{2^{m+1}} - 1 &= \left(k^{2^m}\right)^2 - 1 = (2^{m+2} \cdot t + 1)^2 - 1 \\ &= 2^{2m+4} t^2 + 2^{m+3} t = 2^{m+3} \left(2^{m+1} t^2 + t\right)\end{aligned}$$

Therefore, $2^{m+3} \mid \left(k^{2^{m+1}} - 1\right)$. Hence, $k^{2^{m+1}} \equiv 1 \pmod{2^{m+3}}$.

Practice 99

Let m and n be positive integers, m be odd, and $(m, 2^n - 1) = 1$. Show that $\sum_{k=1}^{m} k^n$ is a multiple of m.

(ref: 4268)

Note that $\{1, 2, \cdots, m\}$ is a complete residue system modulo m.

Meanwhile, because m is odd, thus $(m, 2) = 1$. This means that $\{2, 4, \cdots, 2m\}$ be a complete residue system modulo m too. This means that

$$\sum_{k=1}^{m} k^n \equiv \sum_{k=1}^{m} (2k)^n \equiv 0 \pmod{m}$$
$$\Longrightarrow \sum_{k=1}^{m} (2k)^n - \sum_{k=1}^{m} (k)^n \equiv (2^n - 1) \sum_{k=1}^{m} k^n \equiv 0 \pmod{m}$$

Because $(m, 2^n - 1) = 1$, therefore we must have $\sum_{k=1}^{m} k^n \equiv 0 \pmod{m}$.

Practice 100

The number obtained from the last two non-zero digits of 90! is equal to n. What is n?

(ref: 1508 - AMC10)

First, the number of trailing zero equals the number of divisor 5 that 90! has:

$$\left\lfloor \frac{90}{5} \right\rfloor + \left\lfloor \frac{90}{5^2} \right\rfloor = 21$$

This means that 90! has 21 trailing zeros. Let $N = \frac{90!}{10^{21}}$. Then the desired answer n equals $N \pmod{100}$.

Clearly, N still has more than two divisors of 2. Hence, $N \equiv 0 \pmod{4}$. In order to calculate $N \pmod{100}$, we just need to compute $N \pmod{25}$.

By the conclusion of practice 22, we know $(5k + 1)(5k + 2)(5k + 3)(5k + 4) \equiv -1 \pmod{25}$. In order to use this conclusion, let

$$M = 1 \times 2 \times 3 \times 4 \times \underline{1} \times 6 \times \cdots \times 86 \times 87 \times 88 \times 89 \times \underline{18}$$

That is, to eliminate all 5 from 90!, i.e. $5 \to 1$, $10 \to 2$, $\cdots$, $25 \to 1$,

..., 90 → 18. Then, we have

$$\begin{aligned}M &= 1\times 2\times 3\times 4\times 1\times 6\times \cdots \times 86\times 87\times 88\times 89\times 18\\ &= (1\times 2\times 3\times 4)(6\times 7\times 8)\cdots(86\times 87\times 88\times 89)\\ &\quad (1\times 2\times 3\times 4)(6\times 7\times 8\times 9)\cdots(16\times 17\times 18)\\ &\quad (1\times 2\times 3)\end{aligned}$$

The 2nd last line above is corresponding to all the numbers which divides 5, but not 25. The last line is corresponding to those numbers which are multiple of 25. Therefore

$$M \equiv (-1)^{10}\times(-1)^3\times(16\times 17\times 19)(1\times 2\times 3) \equiv 24 \pmod{25}$$

Meanwhile, we have $2^{21} \equiv (2^{10})^2 \times 2 \equiv (-1)^2 \times 2 \equiv 2 \pmod{25}$. It follows that

$$N = \frac{M}{2^{21}} \equiv \frac{24}{2} \equiv 12 \pmod{25}$$

Together with the fact of $N \equiv 0 \pmod 4$, we found $N \equiv \boxed{12} \pmod{100}$

Practice 101

Find all ordered integer pairs (x, y) such that $x^3 + y^3 = 2019$.

(ref: 4249)

Let n be an integer, then

$$n \equiv 0, \pm 1, \pm 2, \pm 3 \pmod 7 \implies n^3 \equiv 0, \pm 1 \pmod 7$$

Therefore, the sum of two cubes can only be

$$x^3 + y^3 \equiv 0, \pm 1, \pm 2 \pmod 7$$

However, $2019 \equiv 3 \pmod 7$. This means the given equation has no solution.

Practice 102

Find the least non-negative residue of 70! (mod 5183).

(ref: 2739)

Because $5183 = 71 \times 73$, let's start by finding the residues of 70! mod 71 and 73.

By Wilson's theorem, $70! \equiv -1 \pmod{71}$.

Next, let $k = 70! \pmod{73}$. Then $71 \times 72 \times k \equiv 70! \times 71 \times 72 \pmod{73}$ $\implies (-2)(-1)k \equiv 72! \pmod{73}$ $\implies 2k \equiv -1 \pmod{73}$.

Note that $2 \times 37 = 74 \equiv 1 \pmod{73}$. So $37 \cdot 2k \equiv 37 \cdot (-1) \pmod{73}$ $\implies k \equiv -37 \equiv 36 \pmod{73}$.

Thus, $70! \equiv -1 \pmod{71}$ and $70! \equiv 36 \pmod{73}$.

Applying CRT to solve these two equations yields: $70! = \boxed{1277}$ mod 5183.

Practice 103

What is the smallest positive integer n such that $20 \equiv n^{15} \pmod{29}$?

(ref: 2614 - PUMaC)

By Fermat's Little Theorem, we have $a^{28} \equiv 1 \pmod{29}$ for all positive integers a which are not multiples of 29.

It follows that $a^{14} \equiv \pm 1 \pmod{29}$, so $a^{15} \equiv \pm a \pmod{29}$ for all such a. Therefore, if $a^{15} \equiv 20 \pmod{29}$, then $\pm a \equiv 20 \pmod{29}$.

The first few positive integers that satisfy $\pm a \equiv 20 \pmod{29}$ are 9, 20, 38, $\cdots$.

We know that $9^{14} = 3^{28} \equiv 1 \pmod{29}$, so $9^{15} \equiv 9 \pmod{29}$.

Next we try $a = 20$, and we find that $20^{14} \equiv (20+29)^{14} \equiv 7^{28} \equiv 1 \pmod{29}$, and so $20^{15} \equiv 20 \pmod{29}$. Therefore, the answer is $\boxed{20}$.

Practice 104

Find one solution to $x^7 \equiv 3 \pmod{11}$.

(ref: 4166)

Any solution x must be co-prime to 11. By Fermat's little theorem, we have $x^{10} \equiv 1 \pmod{11}$. By Euclid's algorithm, we have

$$3 \times 7 - 2 \times 10 = 1$$

It follows that

$$3^3 \equiv \left(x^7\right)^3 \equiv \left(x^{10}\right)^2 \cdot x \equiv x \pmod{11}$$

Hence we find one solution $x \equiv 3^3 \equiv \boxed{5} \pmod{11}$.

Practice 105

Solve
$$\begin{cases} x &\equiv 2 & \pmod{3} \\ x &\equiv 2 & \pmod{5} \\ x &\equiv -3 & \pmod{7} \\ x &\equiv -2 & \pmod{13} \end{cases}$$

(ref: 2639)

This system can be solved by the Chinese Remainder Theorem (CRT).

$$m = 3 \times 5 \times 7 \times 13 = 1365$$

and,
$$M_1 = 455, M_2 = 273, M_3 = 195, M_4 = 105$$
accordingly,
$$M_1^{-1} = -1, M_2^{-1} = 2, M_3^{-1} = -1, M_4^{-1} = -1$$
Therefore,
$$\begin{aligned} x &= 455 \times (-1) \times 2 + 273 \times 2 \times 2 + 195 \times (-1) \times (-3) + \\ & \quad 105 \times (-1) \times (-2) \\ &= \boxed{557 \pmod{1365}} \end{aligned}$$

Practice 106

Solve
$$\begin{cases} 4x \equiv 14 \pmod{15} \\ 9x \equiv 11 \pmod{20} \end{cases}$$

(ref: 2662)

Because $gcm(15, 20) \neq 1$, CRT cannot be applied directly. As such, we decompose the relation first. The given equations are equivalent to:
$$\begin{cases} 4x \equiv 14 \pmod{3} \\ 4x \equiv 14 \pmod{5} \\ 9x \equiv 11 \pmod{4} \\ 9x \equiv 11 \pmod{5} \end{cases}$$

This system can be simplified as
$$\begin{cases} x \equiv 2 \pmod{3} \\ x \equiv -1 \pmod{5} \\ x \equiv 3 \pmod{4} \\ x \equiv 1 \pmod{5} \end{cases}$$

The 2st and the 4th equations contradict to each other. Hence, this system does not have any solution.

Practice 107

Let n be a positive integer and function $S_1(n)$ return the square of the sum of n's digits. Additionally, let $S_{k+1}(n) = S_1(S_k(n))$, where k is a positive integer. Find the value of $S_{1991}(2^{1990})$.

(ref: 4245 - China)

Let $N = 2^{1990}$. First, because $2^{1990} < 8^{664} < 10^{664}$, N has at most 664 digits. Therefore

$$S_1(N) < (9 \times 664)^2 < 4 \times 10^7$$

Using a similar reasoning, we have

$$S_2(N) \leq (3 + 9 \times 7)^2 < 4400$$
$$S_3(N) \leq (3 + 9 \times 3)^2 < 900 = 30^2$$

By definition, values of all $S_{k+1}(n)$ are square numbers whose square root is the sum of the digits of $S_k(n)$. Let $S_2(N) = a$. Then we have $S_3(N) = a^2$ and $a < 30$. With the MOD by 9 technique, we have

$$a \equiv (S_1(N))^2 \equiv N^4 \equiv 8^{2653} \times 2 \equiv -2 \equiv 7 \pmod{9}$$

Because $0 < a < 30$, $a = 7$, 16, or 25. Therefore, $S_3(N) = 49$, 256, or 625. And, $S_4(N) = 169$, $S_5(N) = 256$, $S_6(N) = 169$.

Now, it is clear that the value of $S_k(N)$ will alternate and $S_{1991}(N) = \boxed{256}$.

Practice 108

Find all the integer pairs (x, y) such that $x^3 = 2^y + 15$.

(ref: 4254)

No such integer pair exists. This can be proved using the MOD method.

It can be shown that for any cube modulo 7 can only result in $x^3 \equiv 0, 1, 6 \pmod{7}$. Meanwhile, any power of 2 modulo 7 can only result in $2^y \equiv 1, 2, 4 \pmod{7}$. Hence $2^y + 15 \equiv 2, 3, 5 \pmod{7}$. Therefore, the two sides cannot be equal.

Practice 109

Show that $x^5 \equiv 3 \pmod{11}$ is not solvable.

(ref: 4165)

Assuming it is solvable, then x must be co-prime to 11. By Fermat's little theorem, we have $x^{10} \equiv 1 \pmod{11}$. However

$$1 \equiv x^{10} \equiv \left(x^5\right)^2 \equiv 3^2 \equiv 9 \pmod{11}$$

cannot hold. This is a contradiction. This means this equation has no solution.

Practice 110

Solve the following relation in integers:

$$x^2 + a^2 = (x+1)^2 + b^2 = (x+2)^2 + c^2 = (x+3)^2 + d^2$$

(ref: 4226)

No solution exists. Note that

$$y^2 \equiv \begin{cases} 0 \pmod{8}, & if\ y \equiv 0 \pmod{4} \\ 1 \pmod{8}, & if\ y \equiv \pm 1 \pmod{4} \\ 4 \pmod{8}, & if\ y \equiv 2 \pmod{4} \end{cases}$$

Then, for any integers y and z, we have

$$y^2 + z^2 \equiv \begin{cases} 0, 1, 4 \pmod{8}, & if\ y \equiv 0 \pmod{4} \\ 1, 2, 5 \pmod{8}, & if\ y \equiv \pm 1 \pmod{4} \\ 0, 4, 5 \pmod{8}, & if\ y \equiv 2 \pmod{4} \end{cases}$$

Now, given x, $x+1$, $x+2$, and $x+3$ form a complete residue system modulo 4. Let's assume

$$\begin{cases} x \equiv 0 \pmod{4}, & \Longrightarrow x^2+a^2 \equiv 0,1,4 \pmod{8} \\ x+1 \equiv 1 \pmod{4}, & \Longrightarrow (x+1)^2+b^2 \equiv 1,2,5 \pmod{8} \\ x+2 \equiv 2 \pmod{4}, & \Longrightarrow (x+2)^2+c^2 \equiv 0,4,5 \pmod{8} \\ x+4 \equiv 3 \pmod{4}, & \Longrightarrow (x+3)^2+d^2 \equiv 1,2,5 \pmod{8} \end{cases}$$

However, the intersection of $\{0,1,4\}$, $\{1,2,5\}$, and $\{0,4,5\}$ is empty. This means no solution will satisfy all these relations. Hence, the given system is insolvable.

Practice 111

Show that there are infinite many composite numbers in the sequence
$$1, 31, 331, 3331, 33331, \cdots$$

(ref: 4212)

Because 31 is a prime, by Fermat's little theorem, we have $10^{30} \equiv 1 \pmod{31}$. Therefore, for any positive inter k, we have $10^{30k} \equiv 1 \pmod{31}$. This means that

$$\frac{1}{3} \times \left(10^{30k} - 1\right) \equiv 0 \pmod{31}$$

Obviously, $(10^{30k} - 1) = \underbrace{33\cdots 3}_{30k}$. It follows that

$$\underbrace{33\cdots 3}_{30k} \times 100 + 31 = \underbrace{33\cdots 3}_{30k} 31$$

is a multiple of 31. Clearly, there are infinite many items in this form in the given sequence.

Practice 112

Let n be a positive integer. Show that there exist n consecutive integers each of which contains a divisor who is a square number greater than 1.

(ref: 4213)

Because there are infinite many prime numbers, it is always possible to select n distinct prime numbers $p_1, p_2, \cdots, p_n$. Consider the following system of congruence

$$\begin{cases} x+1 \equiv 0 \pmod{p_1^2} \implies x \equiv -1 \pmod{p_1^2} \\ x+2 \equiv 0 \pmod{p_2^2} \implies x \equiv -2 \pmod{p_2^2} \\ \cdots \\ x+n \equiv 0 \pmod{p_n^2} \implies x \equiv -n \pmod{p_n^2} \end{cases}$$

Because all the modulo are pairwise co-prime, by the Chinese Remainder Theorem, this system is solvable. Therefore, there exist n consecutive numbers $(x+i)$, $(1 \le i \le n)$, where $p_i^2 \mid x+i$.

Practice 113

Let a, b, and x_0 all be positive integers. Sequence $\{x_n\}$ is defined as $x_{n+1} = ax_n + b$ where $n \ge 1$. Show that $x_1, x_2, \cdots$ cannot be all prime.

(ref: 4257)

If x_1 is not a prime, then the claim holds. Otherwise, suppose $x_1 = p$ is a prime. Then $p > a$. Meanwhile, at least two of the next $(p+1)$ numbers, $x_2, x_3, \cdots, x_{p+2}$, are congruent to each other modulo p. Suppose $x_m \equiv x_n \pmod{m}$ where integers $m > n > 1$. Then we have

$$x_m - x_n \equiv (ax_{m-1} + b) - (ax_{n-1} + b) \equiv a(x_{m-1} - x_{n-1}) \equiv 0$$

$$\therefore x_{m-1} - x_{n-1} \equiv 0 \pmod{p}$$

Continue this substitution will eventually lead to

$$x_{m-n+1} - x_1 \equiv 0 \pmod{p}$$

This means $p \mid x_{m-n+1}$. Meanwhile, $\{x_n\}$ is a strictly increasing sequence because both a and b are positive integers. Therefore $x_{m-n+1} > x_1 = p$. Hence we conclude x_{m-n+1} is a composite.

Practice 114

Let n be an integer which is divisible by neither 2 nor 5. Show that n must be divisible by a number whose digits are all 1.

(ref: 4260)

Consider the following $(n+1)$ integers.

$$1, 11, \cdots, \underbrace{11\cdots 11}_{n+1}$$

If any of them is a multiple of n, then the claim holds. Otherwise, at least two of them must be congruent to each other modulo n. Let these two numbers be $\underbrace{11\cdots 11}_{m}$ and $\underbrace{11\cdots 11}_{k}$ where integers $m > k$. Then we have

$$n \mid \underbrace{11\cdots 11}_{m} - \underbrace{11\cdots 11}_{k} = \underbrace{11\cdots 11}_{m-k}\underbrace{00\cdots 00}_{k}$$

We know that n is a multiple of neither 2 nor 5. Therefore

$$n \mid \underbrace{11\cdots 11}_{m-k}\underbrace{00\cdots 00}_{k} \implies n \mid \underbrace{11\cdots 11}_{m-k}$$

Practice 115

Find the smallest integer N such that $\varphi(n) \geq 5$ holds for all integer $n \geq N$.

(ref: 4194)

Trying a few small numbers finds that it appears $\boxed{13}$ could be a candidate. Let's prove this.

Let's write n in the form of $2^k \cdot m$ where k is a non-negative integer and m is an odd integer.

If $k \geq 4$, then

$$\varphi(n) = \varphi(2^k)\varphi(m) \geq \varphi(2^k) = 2^{k-1} \geq 8$$

Therefore, it is sufficient to just further investigate those cases where $k = 0, 1, 2, 3$.

When $k = 0$, then n is an odd.

- If n is a prime, that $\varphi(n) = n - 1 \geq 13 - 1 = 12$.
- If n divides two distinct prime numbers $p > q$, then we must have $p \geq 5$ and $q \geq 3$. Hence,

$$\varphi(n) \geq \varphi(pq) = \varphi(p)\varphi(q) = (p-1)(q-1) \geq 8$$

- If n divides a square of prime number p, then $p \geq 3$. It follows that

$$\varphi(n) \geq \varphi(p^2) = p(p-1) \geq 6$$

This case also means that for any odd number n, $\varphi(n) \geq \min(n - 1, 6)$. We will reuse this conclusion.

When $k = 1$, then $n = 2m$ where m is an odd integer greater than $\lfloor 13/2 \rfloor = 6$. Then we have

$$\varphi(n) = \varphi(2)\varphi(m) = \varphi(m) \geq \min(m - 1, 6) \geq 6$$

When $k = 2$, then $n = 4m$ where m is odd and $m > \lfloor 13/4 \rfloor = 3$. Therefore

$$\varphi(n) = \varphi(4)\varphi(m) = 2\varphi(m) \geq 2 \times \min(m - 1, 6) \geq 2 \times 3 = 6$$

When $k = 3$, then $n = 8m$ where $m > \lfloor 17/8 \rfloor = 2$. Because m is odd, we find $m \geq 3$. Therefore

$$\varphi(n) = \varphi(8)\varphi(m) = 4\varphi(m) \geq 4 \times \min(m-1, 6) \geq 4 \times 2 = 8$$

Hence, we conclude 13 is the desired answer.

Practice 116

Find all prime number p such that both $(4p^2 + 1)$ and $(6p^2 + 1)$ are prime numbers.

(ref: 174 - Poland)

When $p = 5$, both $4p^2 + 1 = 101$ and $6p^2 + 1 = 151$ are prime. Therefore $p = 5$ is one solution. We are going to show that this is the only solution.

When $p \equiv \pm 1 \pmod 5$, we have $4p^2 + 1 \equiv 0 \pmod 5$. This means that $(4p^2 + 1)$ is a multiple of 5 which cannot be prime.

Meanwhile, when $p \equiv \pm 2 \pmod 5$, we have $6p^2 + 1 \equiv 0 \pmod 5$. This means that $(6p^2 + 1)$ is a multiple of 5 which is not prime.

Practice 117

Ms. Math's kindergarten class has 16 registered students. The classroom has a very large number, N, of play blocks which satisfies the conditions:

- If 16, 15, or 14 students are present in the class, then in each case all the blocks can be distributed in equal numbers to each student, and

- There are three integers $0 < x < y < z < 14$ such that when x, y, or z students are present and the blocks are distributed in equal numbers to each student, there are exactly three blocks left over.

Find the sum of the distinct prime divisors of the least possible value of N satisfying the above conditions.

(ref: 190 - AIME)

N must be a common multiple of 14, 15, and 16. Let k be their least common multiple, i.e. $k = 2^4 \cdot 3 \cdot 5 \cdot 7$. Then, $N = nk$ where n is a positive integer.

Because 1, 2, 3, 4, 5, 6, 7, 8, 10, and 12 all divide k, so $x, y, z = 9, 11, 13$. Then we have the following three modulo equations:

$$\begin{cases} nk \equiv 3 \pmod{9} \\ nk \equiv 3 \pmod{11} \\ nk \equiv 3 \pmod{13} \end{cases}$$

This means that nk must be three more than a common multiple of 9, 11, and 13, or $nk = 9 \times 11 \times 13 \times p + 3$ where p is an integer. Meanwhile, it must be divisible by k. Therefore there exists an integer q such that

$$9 \times 11 \times 13 \times p + 3 = (2^4 \times 3 \times 5 \times 7 \times 1) \times q$$
$$429 \times p + 1 = 560 \times q$$

This is a classic linear indeterminate equation. The least positive integer solution is $(p, q) = (171, 131)$. Setting $p = 171$ into $(9 \times 11 \times 13 \times p + 3)$ and factorizing find the distinct prime divisors are 2, 3, 5, 7, and 131. Their sum is $\boxed{148}$.

Practice 118

Let $\mathcal{S}$ be the set of all perfect squares whose rightmost three digits in base 10 are 256. Let $\mathcal{T}$ be the set of all numbers of the form $\frac{x-256}{1000}$, where x is in $\mathcal{S}$. In other words, $\mathcal{T}$ is the set of numbers when the last three digits of each number in $\mathcal{S}$ are truncated. Find the remainder when the tenth smallest element of $\mathcal{T}$ is divided by 1000.

(ref: 219 - AIME)

Firstly, we note that if x^2 ends with 256, then $(x + 1000)^2$ must end with 256 too because $(x + 1000)^2 \equiv x^2 \pmod{1000}$. Hence, we just need to focus on numbers less than 1000.

Next, x^2 ends with 256 is equivalent to say $x^2 - 256 = 1000 \cdot k$ where k is an integer. This implies $(x + 16)(x - 16)$ is a multiple of $10000 = 2^3 \times 5^3$.

Now because $x + 16 \not\equiv x - 16 \pmod{5}$, therefore $(x + 16)$ and $(x - 16)$ cannot both be multiples of 5. Hence, one of them must be a multiple of $5^3 = 125$.

Meanwhile, because $x + 16 \equiv x - 16 \pmod 4$, therefore both of them must be multiples of 4. Otherwise, if neither of them is a multiple of 4, then $(x + 16)(x - 16)$ cannot be a multiple of $2^3 = 8$.

Combining both facts means one of $(x+16)$ and $(x-16)$ is a multiple of $125 \times 4 = 500$ or $x = 500n \pm 16$. Now, it is just a matter to check 16, $(500 - 16)$, $(500 + 16)$, and $(1000 - 16)$. Squares of all of them end with 256.

Therefore, the 10th smallest qualifying number is $(2000+500-16) =$

2484. And the desired answer is

$$\frac{2484^2 - 256}{1000} \equiv \boxed{170} \pmod{1000}$$

Practice 119

For a positive integer p, define the positive integer n to be p-safe if n differs in absolute value by more than 2 from all multiples of p. For example, the set of 10-safe numbers is $\{3, 4, 5, 6, 7, 13, 14, 15, 16, 17, 23, \ldots\}$. Find the number of positive integers less than or equal to $10,000$ which are simultaneously 7-safe, 11-safe, and 13-safe.

(ref: 236 - AIME)

For any integer n, let $r = n \pmod p$ where $0 \leq r < p$. Then n is p-safe if and only if $p - 2 > r > 2$. Therefore, there are $p - 5$ qualified r.

Therefore a number n that meets the requirement of this problem can have 2 different residues (mod 7), 6 different residues (mod 11), and 8 different residues (mod 13). Because 7, 11, and 13 are pairwise co-prime, we assert these exist one unique solution to a combination of one each of these congruent relations by the Chinese remainder theorem within the range of 1 to $7 \times 11 \times 13 = 1001$. Therefore, there are totally $2 \times 6 \times 8 = 96$ solutions not exceeding 1001. This means 960 solutions not exceeding 10010.

Meanwhile, it is easy to check manually and find 10006 and 10007 are two solutions greater than 10000. Excluding these two, we find the final answer to this question is $\boxed{958}$.

Practice 120

Let R be the set of all possible remainders when a number of the form 2^n, where n is a non-negative integer, is divided by 1000. Let S be the sum of the elements in R. Find the remainder when S is divided by 1000.

(ref: 250 - AIME)

We want to find two distinct integer $i < j$ such that $2^i \equiv 2^j$ (mod 1000). If so, then it is sufficient to find the remainders of $2^0, 2^1, 2^2, \cdots, 2^{j-1}$ MOD 1000 in order to get the answer.

The condition $2^i \equiv 2^j$ (mod 1000) is equivalent to $2^i \equiv 2^j$ (mod 8) and $2^i \equiv 2^j$ (mod 125).

For MOD 8, we note that for any integer $k \geq 3$, it always hold that $2^k \equiv 0$ (mod 8).

By practice 43, we know the multiplicative order of 2 modulo 125 is 100. Therefore we can conclude that $2^1, 2^2, \cdots, 2^{100}$ should all have different residues modulo 125. This is because if there exist positive integer $m < n \leq 100$ such that $2^m \equiv 2^n$ (mod 125), we will have $2^{n-m} \equiv 1$ (mod 125) and $n - m < 100$.

Combining these two results gives $2^3 \equiv 2^{103}$ (mod 1000). It follows that

$$S \equiv 2^0 + 2^1 + \cdots + 2^{102} \equiv 2^{103} - 1 \equiv 2^3 - 1 \equiv \boxed{7} \pmod{1000}$$

Practice 121

The number 2017 is prime. Let $S = \sum_{k=0}^{62} \binom{2014}{k}$. What is the remainder when S is divided by 2017?

(ref: 474 - AMC12)

First, let's simplify C_{2014}^k as

$$C_{2014}^k \equiv \frac{2014 \times 2013 \times \cdots (2014-k+1)}{k!}$$
$$\equiv \frac{(-3)(-4)\cdots(-k-2)}{k!}$$
$$\equiv (-1)^k C_{k+2}^k$$
$$\equiv (-1)^k C_{k+2}^2$$
$$\equiv (-1)^k \times \frac{(k+2)(k+1)}{2} \quad (\bmod\ 2017)$$

When k is even, let $k = 2m$. Then the sum of those "even" terms in the above expression equals

$$\sum_{m=0}^{31} \frac{(2m+2)(2m+1)}{2} = \sum_{m=0}^{30}(2m^2+3m+1)$$

When k is odd, let $k = 2m+1$. Then the sum of those "odd" terms above equals

$$\sum_{m=0}^{30}(-1) \times \frac{(2m+3)(2m+2)}{2} = -\sum_{m=0}^{30}(2m^2+5n+3)$$

Adding them together we have

$$S = 2 \times 31^2 - 2\sum_{m=0}^{30} m + 3 \times 31 + 32 - 3 \times 31 = \boxed{1024}$$

Practice 122

Seven students count from 1 to 1000 as follows:

- Alice says all the numbers, except she skips the middle number in each consecutive group of three numbers. That is, Alice says 1, 3, 4, 6, 7, 9, . . ., 997, 999, 1000.

- Barbara says all of the numbers that Alice doesn't say, except she also skips the middle number in each consecutive group of three numbers.

- Candice says all of the numbers that neither Alice nor Barbara says, except she also skips the middle number in each consecutive group of three numbers.

- Debbie, Eliza, and Fatima say all of the numbers that none of the students with the first names beginning before theirs in the alphabet say, except each also skips the middle number in each of her consecutive groups of three numbers.

- Finally, George says the only number that no one else says.

What number does George say?

(ref: 1457 - AMC10)

Such a puzzle is clearly related to modular arithmetics. Let's try to find the pattern.

- Alice skips all the numbers n which satisfies $n \equiv 2 \pmod{3}$. Her group contains 3 numbers. Each group produces one number for Barbara to process.

- Barbara's group contains 3 numbers provided by Alice which is equivalent to $3^2 = 9$ original numbers. It is easy to find out that all the number Barbara skips are $n \equiv (3+2) \pmod{3^2}$ or $n \equiv 5 \pmod 9$

- Similarly, Candice's group will contain $3^3 = 27$ original numbers and skip those $n \equiv (9+5) \pmod{3^3}$ or $n \equiv 14 \pmod{27}$
- Debbie: $n \equiv 27 + 41 \pmod{3^4}$ or $n \equiv 41 \pmod{81}$
- Eliza: $n \equiv 81 + 41 \pmod{3^5}$ or $n \equiv 122 \pmod{243}$
- Fatima: $n \equiv 243 + 122 \pmod{3^6}$ or $n \equiv 365 \pmod{729}$

There is only number n not exceeding 1000 which satifies $n \equiv 365 \pmod{729}$ which is $\boxed{365}$.

Practice 123

N delegates attend a round-table meeting, where N is an even number. After a break, these delegates randomly pick a seat to sit down again to continue the meeting. Prove that there must exist two delegates so that the number of people sitting between them is the same before and after the break.

(ref: 2088)

Fix a seat, and then label these delegates from 0 to $N-1$ clockwise based on their seats. After the break, label these delegates' seat as $a_0, a_1, \cdots, a_{n-1}$.

Now, let's consider $b_i = (a_i - i) \pmod n$, where $i = 0, 1, \cdots, N-1$.

If there exist m and k, such that $b_k = a_m$, then $a_k - a_m = k - m$. This means the intervals between the kth and the mth delegates keeps the same.

Otherwise, if all b_i's are different, they must take distinct values from $\{0, 1, 2, \cdots, N-1\}$. However this is impossible because, on one hand, $\sum_{i=0}^{N-1} b_i = \frac{N(N-1)}{2} \equiv \frac{N}{2}$ (mod N). On the other hand, the result should be 0 by definition because $\sum_{i=0}^{N-1} b_i = \sum_{i=0}^{N-1} (a_i - i) = 0$.

Practice 124

Prove that if p and $(p^2 + 8)$ are prime, then $(p^3 + 8p + 2)$ is prime.

(ref: 2217)

First, when $p = 2$, $p^2 + 8 = 12$ is not a prime. Next, when $p = 3$, both $p^2 + 8 = 17$ and $p^3 + 8p + 2 = 53$ are prime.

Therefore, the conclusion holds when $p \leq 3$. Now, let's show that when $p > 3$, $(p^2 + 8)$ cannot be a prime.

In this case, given p is prime, we have have $p \not\equiv 0$ (mod 3). However,

$p^2 + 8 \not\equiv 0$ (mod 3) $\implies p^2 \not\equiv 1$ (mod 3) $\implies p \not\equiv \pm 1$ (mod 3)

This is impossible because one of $p \equiv 0$ (mod 3) and $p \equiv \pm 1$ (mod 3) must hold.

Practice 125

Given that there are 24 primes between 3 and 100, inclusive, what is the number of ordered pairs (p, a) with p prime, $3 \leq p < 100$, and $1 \leq a < p$ such that the sum $a + a^2 + a^3 + \cdots + a^{(p-2)!}$ is not divisible by p?

(ref: 2615 - PUMaC)

If $a = 1$, then the sum just becomes $(p-2)!$, which is never divisible by p. So since there are 24 odd primes between 2 and 100, there are 24 solutions of the form $(p, 1)$.

Next, suppose $a \neq 1$. The sum can then be written as

$$a + a^2 + \cdots + a^{(p-2)!} = a \times \frac{a^{(p-2)!} - 1}{a - 1} = \frac{a}{a-1} \cdot \left(a^{(p-2)!} - 1\right)$$

Since $1 < a < p$, the term $\frac{a}{(a-1)}$ does not contribute to whether the sum is divisible by p. So it sufficient to consider the term $a^{(p-2)!} - 1$. Now look at the following cases.

- If $p = 3$, then the sum is just a which is not divisible by p. So $(3, 2)$ is a valid solution.

- If $p = 5$, then the sum is

  $$\frac{a}{a-1} \cdot (a^6 - 1) \equiv \frac{a}{a-1} \cdot (a^2 - 1) \equiv a(a+1) \pmod{5}$$

  by Fermat's Little Theorem. Plugging in $a = 2, 3, 4$ shows that $(5, 2)$ and $(5, 3)$ are the only solutions here.

- If $p > 5$, then $2 \neq (p-1)/2$. Moreover, we also have $2 \mid (p-2)!$ and $(p-1)/2 \mid (p-2)!/2$ since $1 < 2, (p-1)/2 < p-2$. Thus $(p-1) \mid (p-2)!$, so by Fermat's Little Theorem $a^{(p-2)!} - 1 \equiv \pmod{p}$. Thus the sum is always divisible by p in this case, and there are no solutions here.

Thus there is a total of $\boxed{27}$ solutions.

Practice 126

If for any integer $k \neq 27$ and $\left(a - k^{2015}\right)$ is divisible by $(27 - k)$, what is the last two digits of a?

(ref: 2621)

Let $f(k) = a - k^{2015}$. Because $(k - 27) \mid f(k)$, we have $f(27) = 0$, i.e. 27 is a root of $f(k)$. This means that $a - 27^{2015} = 0$ which implies that the last two digits of a is the same as those of 27^{2015}.

$$27^{2015} = 3^{6045} = (3^4)^{1511} \times 3 = 81^{1511} \times 3$$

Now by applying the "quick way to find the tens digit" trick, we know the tens digit of 81^{1511} is 8. Meanwhile, its units digit obviously is 1. Hence, 81^{1511} ends with 81 which leads to 27^{2015} ends with $\boxed{43}$.

Practice 127

A positive integer n is said to be good if there exists a perfect square whose sum of digits in base 10 is equal to n. For instance, 13 is good because $7^2 = 49$ and $4 + 9 = 13$. How many good numbers are among $1, 2, 3, \cdots, 2007$?

(ref: 2816)

If a positive integer is a multiple of 3, then its square is a multiple of 9, and so is the sum of the digits of its square.

If a positive integer is not a multiple of 3, then its square is 1 more than a multiple of 3, and so is the sum of the digits of its square.

Therefore, we only need to study two categories of candidates.

The square of $\underbrace{9\cdots 9}_{m}$ is $\underbrace{9\cdots 9}_{m-1}8\underbrace{0\cdots 0}_{m-1}1$. Its digit sum is $9m$. Therefore, all multiples of 9 are good. There are $2007 \div 9 = 223$ of them not exceeding 2007.

The square of $\underbrace{3\cdots 3}_{m}5$ is $\underbrace{1\cdots 1}_{m}2\underbrace{2\cdots 2}_{m+1}5$ Its digit sum is $3m+7$. Since 1 and 4 are also good, all numbers 1 more than a multiple of 3 are good, and there are $2007 \div 3 = 669$ of them.

Hence there are altogether $223 + 669 = \boxed{992}$ good numbers not exceeding 2007.

Practice 128

Let x be an integer and p is a prime divisor of $(x^6+x^5+\cdots+1)$. Show that $p=7$ or $p\equiv 1\pmod 7$.

(ref: 3795)

Obviously, if $x=1$, then $p=7$.

When $x\neq 1$, then p divides $\frac{x^7-1}{x-1}$. Hence, $x^7\equiv 1\pmod p$. This implies $p\nmid x$. Then, by Fermat's little theorem, we have $x^{p-1}\equiv 1\pmod p$. This is followed by

$$x^{(7,p-1)}\equiv 1\pmod p$$

where $(7,p-1)$ is the greatest common divisor of 7 and $(p-1)$.

If $p\not\equiv 1\pmod 7$, i.e. $7\nmid (p-1)$, then $(7,p-1)=1$. This will lead to $x\equiv 1\pmod p$ by the equation above. Then,

$$x^6+x^5+\cdots+1\equiv 1^6+1^6+\cdots+1^6\equiv 7\pmod p$$

and

$$p\mid (x^6+x^5+\cdots+1)\implies x^6+x^5+\cdots+1\equiv 0\pmod p$$

This will lead to the conclusion $p=7$. Hence, we find the claim holds.

Practice 129

Let p be an odd prime number. For positive integer k satisfying $1\le k\le p-1$, the number of divisors of $kp+1$ between k and p exclusive is a_k. Find the value of $a_1+a_2+\ldots+a_{p-1}$.

(ref: 3841 - Japan)

Let's first examine a few simplest cases and try to find some clues from these trials.

When $p = 3$:

$p = 3$	k	$kp + 1$	a_i	divisors
	1	4	1	(2)
	2	7	—	—
	$a_1 + a_2 =$		1	(2)

When $p = 5$:

$p = 5$	k	$kp + 1$	a_i	divisors
	1	6	2	(2, 3)
	2	11	—	—
	3	16	1	(4)
	4	21	—	—
	$a_1 + \cdots + a_4$		3	(2, 3, 4)

When $p = 7$:

$p = 7$	k	$kp + 1$	a_i	divisors
	1	8	2	(2, 4)
	2	15	2	(3, 5)
	3	22	—	—
	4	29	—	—
	5	36	1	(6)
	6	43	—	—
	$a_1 + \cdots + a_6$		5	(2, 3, 4, 5, 6)

It appears that

- The answer is $\boxed{p - 2}$,
- Every number in $2, 3, \cdots, p - 1$ appears once and only once among the divisors, and
- The largest $kp + 1$ (i.e. when $k = p - 1$) does not have any qualified divisor.

Hence, a hint for solve this problem app

We note that there are totally $(p-2)$ divisors $2, 3, \cdots, p-1$, and there are totally $(p-2)$ numbers in the form of $(kp+1)$, excluding the largest one. Therefore, the observation listed above seems to indicate that the original problem is equivalent to showing that every number in $2, 3, \cdots, p-1$ divides on and only one number in the following list:

$$p+1, 2p+1, 3p+1, \cdots, (p-2)p+1$$

This claim can go further.

Claim: Fixed $m (1 < m < p)$, then m contribute exactly once in one of $a_1, a_2, \ldots, a_{m-1}$.

Proof. Consider the following $m-1$ numbers:

$$p+1, 2p+1, \ldots, (m-1)p+1,$$

since $\gcd(m, p) = 1$, and none of them equal to 1 modulo m, so the statement follows.

Hence the answer is $\boxed{p-2}$.

Practice 130

Compute $3^{3^{3^{\cdots^3}}}$ (mod 100), where there are 2012 times.

(ref: 4162)

By Euler's theorem, we have $3^{40} \equiv 1 \pmod{100}$ because $\varphi(100) = 40$. Let's compute $3^{3^{3^{\cdots^3}}}$ (mod 40), with 2011 times. Recursively, we need to compute $3^{3^{\cdots^3}}$ (mod 16), with 2010 times, $3^{3^{\cdots^3}}$ (mod 8), with 2009 times, $3^{3^{\cdots^3}}$ (mod 4), with 2008 times, and $3^{3^{\cdots^3}}$ (mod 2), with 2007 times. The last expression obviously equals 1. Therefore,

$$\underbrace{3^{3^{3^{3^{\cdots 3}}}}}_{2008\ times} \equiv 3 \pmod{4}$$

$$\implies \underbrace{3^{3^{3^{3^{\cdots 3}}}}}_{2009\ times} \equiv 3^3 \equiv 3 \pmod{8}$$

$$\implies \underbrace{3^{3^{3^{3^{\cdots 3}}}}}_{2010\ times} \equiv 3^3 \equiv 11 \pmod{16}$$

$$\implies \underbrace{3^{3^{3^{3^{\cdots 3}}}}}_{2011\ times} \equiv 3^{11} \equiv (3^4)^2 \times 3^3 \equiv 27 \pmod{40}$$

$$\implies \underbrace{3^{3^{3^{3^{\cdots 3}}}}}_{2012\ times} \equiv 3^{27} \equiv \boxed{87} \pmod{100}$$

The last step can utilize "the quick way to find the tens digit" technique to complete.

Practice 131

How many positive integers not exceeding 100 are there such that the value of $(3^x - x^2)$ is a multiple of 5?

(ref: 4176)

Because $3^4 \equiv 1 \pmod{5}$, therefore $3^{x+4} \equiv 3^x \pmod{5}$. Meanwhile, it is obvious that $x + 5 \equiv x \pmod{5}$. It follows that

$$3^{(x+20)} + (x+20)^2 \equiv 3^x + x^2 \pmod{5}$$

This means that we only need to examine positive integers not exceeding 20 to get the result. There are only four possible values 2, 4, 16, and 18 satisfying the requirement. Therefore the answer to the given question is $4 \times 5 = \boxed{20}$.

Practice 132

Let $\mathbb{S}$ be the set of integers between 1 and 2^{40} that contain two 1s when written in base 2. What is the probability that a random integer from $\mathbb{S}$ is divisible by 9?

(ref: 4177 - AIME)

Because $2^6 \equiv 1 \pmod 9$, we find $2^{n+6} \equiv 2^n \pmod 9$ holds for any integer n. This means that power of 2 forms a cyclic of length 6 in MOD 9: 2, 4, -1, -2, -4, and 1.

Any number that contains two 1's when written in base 2 must be in the form of $(2^a + 2^b)$ when written in base 10 where $a \neq b$. In order to make it a multiple of 9, or $2^a + 2^b \equiv 0 \pmod 9$, a and b must be in the following pairs

$$(6k, 6m+3), (6k+1, 6m+4), (6k+2, 6m+5)$$

where k and m are two integers.

Within the given range, $1 = 2^0$ and 2^{40}, there are $7 \times 7 = 49$ choices for the first pair, $7 \times 6 = 42$ choices for the second pair and $7 \times 6 = 42$ choices for the third pair. Therefore there are totally $49 + 42 + 42 = 133$ qualified numbers. Meanwhile, there are totally $C_{40}^2 = 780$ numbers that have exactly two 1s in their binary representations. Thus the final answer is $\boxed{\dfrac{133}{780}}$.

Practice 133

(Thue's theorem) Let p be a prime. Show that for any integer a such that $p \nmid a$, there exist positive integers x, y not exceeding $\lfloor \sqrt{p} \rfloor$ satisfying $ax \equiv y \pmod p$ or $ax \equiv -y \pmod p$.

(ref: 4186)

First, consider choosing x and y from $0, 1, \cdots, \lfloor\sqrt{p}\rfloor$. There are $\left(\lfloor\sqrt{p}\rfloor + 1\right)^2$ possible pairs. The number is more than p. Therefore, as (x, y) varies over these possible combinations, there will be more than p expressions in the form of $(ax - y)$. By the pigeonhole principle, there must be at least two such expressions are congruent to each other modulo p. Let them be $(x_1, y_1) \neq (x_2, y_2)$, but $ax_1 - y_1 \equiv ax_2 - y_2 \pmod{p}$. This implies $a(x_1 - x_2) \equiv (y_1 - y_2) \pmod{p}$. Let $x = |x_1 - x_2|$ and $y = |y_1 - y_2|$, we will have $ax \equiv \pm y \pmod{p}$.

All left to do is to exclude the possibility that $x = 0$ or $y = 0$. If so, both x and y will be in the required range.

If $x = 0$, then $x_1 = x_2$ and $a(x_1 - x_2) = 0 \equiv y_1 - y_2 \pmod{p}$. Because $0 \leq y_1, y_2 < p$, then it must hold that $y_1 = y_2$. This contradicts to the assumption of $(x_1, y_1) \neq (y_1, y_2)$.

Meanwhile, if $y = 0$, then $y_1 = y_2$ and $a(x_1 - x_2) \equiv 0 \pmod{p}$. Because p is a prime and $p \nmid a$, this means that $x_1 - x_2 \equiv 0 \pmod{p}$. By applying the similar logic stated previously, this will also lead to the same contradiction.

Practice 134

Let p be a prime. Show that there exist infinitely many positive integer n such that $p \mid (2^n - n)$.

(ref: 4205)

When $p = 2$, then any even number n will satisfy the requirement.

When $p > 2$, by Fermat little theorem, we have $2^{p-1} \equiv 1 \pmod{p}$. This means that we have $2^{k(p-1)} \equiv 1 \pmod{p}$ for any integer k.

Now, if we can show that there exist infinitely many k such that $k(p-1) \equiv 1 \pmod{p}$, then letting $n = k(p-1)$ will yield

$$2^n \equiv 1 \equiv n \pmod{p} \implies p \mid (2^n - n)$$

Indeed, there exist infinitely many such k because as long as $k \equiv -1$ (mod p), we will have $k(p-1) \equiv 1$ (mod p).

Practice 135

Let p be a prime number and $\lfloor x \rfloor$ denote the largest integer not exceeding real number x. Show that

$$C_n^p \equiv \left\lfloor \frac{n}{p} \right\rfloor \pmod{p}$$

(ref: 4222)

Let i be the least non-negative integer satisfying $i \equiv n \pmod{p}$. Then, we have

$$\left\lfloor \frac{n}{p} \right\rfloor = \frac{n-i}{p} \quad (0 \le i < p)$$

We also note because $n \equiv i \pmod{p}$, it must hold that

$$\begin{aligned} n-1 &\equiv i-1 & \pmod{p} \\ n-2 &\equiv i-2 & \pmod{p} \\ &\cdots \\ n-i+1 &\equiv 1 & \pmod{p} \\ n-i-1 &\equiv -1 \equiv p-1 & \pmod{p} \\ n-i-2 &\equiv -2 \equiv p-2 & \pmod{p} \\ &\cdots \\ n-p+1 &\equiv i+1 & \pmod{p} \end{aligned}$$

It follows that

$$\begin{aligned} & n(n-1)\ldots(n-i+1)(n-i-1)\ldots(n-p+1) \\ \equiv\ & i(i-1)\ldots 1 \cdot (p-1)\ldots(i+1) \\ \equiv\ & (p-1)! \pmod{p} \end{aligned}$$

Therefore,

$$C_n^p = \frac{n(n-1)\ldots(n-i+1)(n-i)(n-i-1)\ldots(n-p+1)}{p!}$$

$$\equiv n(n-1)\ldots(n-i+1)(n-i-1)\ldots(n-p+1)\frac{n-i}{p!}$$
$$\equiv (p-1)!\frac{n-i}{p!}$$
$$\equiv \frac{n-i}{p}$$
$$\equiv \left\lfloor \frac{n}{p} \right\rfloor \pmod{p}$$

Practice 136

Let integer $N = \lfloor (\sqrt{29} + \sqrt{21})^{2020} \rfloor$ where $\lfloor x \rfloor$ denotes the largest integer not exceeding x. Find the last two digits of N.

(ref: 4242)

Let $A = \sqrt{29}+\sqrt{21}$ and $B = \sqrt{29}-\sqrt{21}$. Then, we have $0 < B < 1$. Meanwhile, by binomial expansion we can show $(A^{2020} + B^{2020})$ is an integer. (see below). Therefore, we have $N = A^{2020} + B^{2020} - 1$.

$$A^{2020} + B^{2020}$$
$$= (\sqrt{29}+\sqrt{21})^{2020} + (\sqrt{29}-\sqrt{21})^{2020}$$
$$= 2 \times \left(C_{2020}^0 (\sqrt{29})^{2020} + C_{2020}^2 (\sqrt{29})^{2018} (\sqrt{21})^2 + \cdots \right.$$
$$\left. + C_{2020}^{2020} (\sqrt{21})^{2020} \right)$$

Clearly, every term on the right is an integer, thus their sum is an integer too.

Meanwhile, we have

$$A^{2020} + B^{2020}$$
$$= (50 + 2\sqrt{29}\sqrt{21})^{1010} + (50 - 2\sqrt{29}\sqrt{21})^{1010}$$
$$= 2 \times \left(C_{1010}^0 \cdot 50^{1010} + \cdots + C_{1010}^2 \cdot 50^2 \cdot (2\sqrt{29}\sqrt{21})^{1008} \right.$$
$$\left. + C_{1010}^{1010} (2 \cdot \sqrt{29}\sqrt{21})^{1010} \right)$$

Every term except the last one is a multiple of 100. Therefore,

$$\begin{aligned} N &= A^{2020} + B^{2020} - 1 \\ &= 2 \cdot C_{1010}^{1010} \left(2 \cdot \sqrt{29}\sqrt{21}\right)^{1010} - 1 \\ &\equiv 2^{1011} \cdot 3^{505} \cdot 7^{505} \cdot 29^{505} - 1 \\ &= \boxed{51} \end{aligned}$$

The last step can utilize the "a quick way to find the tens digit" techniques

Practice 137

Let m and n be two positive integers, find the minimal value of $\mid 12^m - 5^n \mid$.

(ref: 4246)

Let $N = \mid 12^m - 5^n \mid$. It is clear that N is an odd number and, neither a multiple of 3 nor a multiple of 5. This means that $N = 1$ or $N \geq 7$.

First, let's show $N \neq 1$.

- If $(12^m - 5^n) = 1$, then taking MOD 4 on both sides yields $-1 \equiv 1 \pmod{4}$. It cannot hold.

- If $(5^n - 12^m) = 1$, then taking MOD 3 on both sides yields $(-1)^n \equiv 1 \pmod{3}$. Therefore n must be even. Taking MOD 5 on the original expression gives $2^m \equiv -1 \pmod{5}$ which means $2^{m+2} \equiv 1 \pmod{5}$. Letting $n = 2k$ and $m+2 = 4l$ will lead to the following relation which is contradicting.

$$5^n - 12^m = 5^{2k} - 12^{4l} = \left(5^k + 12^{2l}\right)\left(5^k - 12^{2l}\right) > 1$$

Therefore, we conclude N cannot equal 1. Hence, $N \geq 7$. Clearly, when $m = n = 1$, we have $N = 7$. This means that the answer is $\boxed{7}$.

Practice 138

Determine all positive integer n such that the following equation is solvable in integers:
$$x^n + (2+x)^n + (2-x)^n = 0$$

(ref: 4258)

Clearly, this equation is insolvable when n is even.

When $n = 1$, there exists one solution $x = 4$.

When n is odd and greater than 1, the equation can be expanded to a polynomial in the following form:
$$x^n + \cdots + 2^{n+1} = 0$$

Rational root theorem asserts that any integral root must be a divisor of 2^{n+1}. When $x > 0$, it is easy to show $x^n + (2+x)^n + (2-x)^n > 0$. Hence, any possible root must be in the form of -2^k where k is a non-negative integer not exceeding $(n+1)$.

- When $k = 0, 1, 2$, it can be verified that none of $x = -1, -2, -4$ is a root.

- When $k \geq 3$, then setting $x = -2^k$ leads to
$$-2^{nk} + (2 - 2^k)^n + (2 + 2^k)^n = 0$$
$$\Leftrightarrow -2^{n(k-1)} + (1 - 2^{k-1})^n + (1 + 2^{k-1})^n = 0$$

Noting $k \geq 3$ and n is a positive integer, the left side of the last equation is congruent to 2 modulo 4. However, the right side $0 \equiv 0 \pmod 4$. Therefore, this equation is insolvable.

In conclusion, the given equation is only solvable when $n = \boxed{1}$.

Practice 139

Let sequence $\{a_n\}$ be $a_n = 2^n + 3^n + 6^n - 1$ where $n \geq 1$. Find the sum of all positive integers which are co-prime to all the a_n.

(ref: 4269)

Clearly, $(1, a_n) = 1$ holds for every n. We are going to show that 1 is the only positive integer satisfying this requirement. If so, the answer is $\boxed{1}$.

Let $m > 1$ be a positive integer which is co-prime to all a_n and p be a prime divisor of m.

First, because $a_2 = 48$, p can be neither 2 nor 3. Otherwise, we will have $(m, a_2) \neq 1$.

If $p > 3$, by Fermat's little theorem, we have $2^{p-1} \equiv 1 \pmod{p}$, $3^{p-1} \equiv 1 \pmod{6}$, and $6^{p-1} \equiv 1 \pmod{p}$.. Then

$$a_{p-2} = 2^{p-2} + 3^{p-2} + 6^{p-2} - 1 \equiv \frac{1}{2} + \frac{1}{3} + \frac{1}{6} - 1 \equiv 0 \pmod{p}$$

This means p divides both m and a_{p-2}, i.e. $(m, a_{p-2}) > 1$.

Hence, we conclude 1 is the only positive integer that is co-prime to all a_n.

Practice 140

Let p be a prime and

$$\frac{a}{b} = \frac{1}{1^2} + \frac{1}{2^2} + \cdots + \frac{1}{(p-1)^2}$$

where a and b are two co-prime positive integers. Show that $p \mid a$.

(ref: 4219)

Multiplying both sides by $((p-1)!)^2$ yields

$$((p-1)!)^2 \cdot \frac{a}{b} = \left(\frac{(p-1)!}{1}\right)^2 + \left(\frac{(p-1)!}{2}\right)^2 + \cdots + \left(\frac{(p-1)!}{p-1}\right)^2$$

Every term on the right side is an integer, therefore the left side must be an integer too.

We are gong to show that the right side is a multiple of p. If so, the left side must be a multiple of p too. However, because p is a prime, we must have $p \nmid (p-1)!$. This will lead to the desired result $p \mid a$.

Because p is a prime, for every integer $k = 1, 2, \cdots, (p-1)$, there exists a multiplicative inverse h such that $kh \equiv 1 \pmod{p}$. Hence, for every k:

$$\frac{(p-1)!}{k} \cdot (hk) \equiv (p-1)! \cdot h \pmod{p}$$

Because $1, 2, \cdots,$ and $(p-1)$ are all distinct, their corresponding inverses are also distinct which means that these inverses are $1, 2, \cdots,$ and $(p-1)$ too.

$$\therefore \sum_{k=1}^{p-1} \left(\frac{(p-1)!}{k}\right)^2$$
$$\equiv \sum_{k=1}^{p-1} \left(\frac{(p-1)!}{k} \cdot (hk)\right)^2$$
$$\equiv \sum_{k=1}^{p-1} ((p-1)! \cdot h)^2$$
$$= ((p-1)!)^2 \sum_{k=1}^{p-1} h^2$$
$$\equiv ((p-1)!)^2 \cdot \frac{1}{6} \cdot (p-1)p(2(p-1)+1)$$
$$\equiv 0 \pmod{p}$$

Practice 141

Let n be an odd integer greater than 3, and $\mathbb{S} = \{0, 1, \cdots, n-1\}$. Show that after removing any element from $\mathbb{S}$, it is always possible to equally divide the remaining elements in $\mathbb{S}$ into two groups such that their sum are congruent modulo n.

(ref: 4210)

Let the element to be removed be $x \in \mathbb{S}$. Then transferring $\mathbb{S} \setminus \{x\}$ to set $\mathbb{T}$ by setting $t_i = (s_i - x) \pmod{n}$ will not change the sum of these element modulo n. This means that it is always possible to *shift* the elements without changing the nature of this problem. Therefore, it is sufficient to prove the claim with an assumption that the removed element is 0.

When $n = 4k + 1$ where k is a positive integer, then we can match the remaining $4k$ element into the following $2k$ pairs:

$$\{1, 4k\}, \{2, 4k-1\}, \cdots, \{2k, 2k+1\}$$

The sum of every pair is $4k + 1 = n$, hence congruent 0 modulo n. Therefore, selecting any k pairs from them to form the first group and the remaining k pairs to form the second group will satisfy the requirement because their sums will be both congruent to 0 modulo n.

When $n = 4k + 3$, where k is a positive integer, then we place $\{1, 2, 4k\}$ into the first group, and $\{3, 4k+1, 4k+2\}$ into the second group. Their sums are both congruent 0 modulo $n = 4k + 3$. Next, matching the remaining elements into the following $(2k - 2)$ pairs:

$$\{4, 4k-1\}, \{5, 4k-2\}, \cdots, \{2k+1, 2k+2\}$$

The sum of every pair equals $(4k + 3)$. Therefore, equally dividing these pairs into the two groups will satisfy the requirement.

Practice 142

Let n be a positive integer not less than 4. Show that there exists a polynomial with integral coefficients

$$f(x) = x^n + a_{n-1}x^{n-1} + a_{n-2}x^{n-2} + \cdots + a_1 x + a_0$$

such that for any positive integer m and any $k \geq 2$ distinct integers $r_1, r_2, \cdots, r_k$, it always hold that $f(m) \neq f(r_1)f(r_2) \cdots f(r_k)$.

(ref: 4211)

Let's construct a polynomial such that for any integer u, it always hold $f(u) \equiv 2 \pmod{4}$. If so, because $k \geq 2$, we always have $f(r_1)f(r_2) \cdots f(r_k) \equiv 0 \pmod{4}$ but $f(m) \equiv 2 \pmod{4}$, therefore they can never equal.

There exist many candidates, the polynomial below is one of them

$$f(x) = (x+1)(x+2)\cdots(x+n) + 2$$

It is clearly a n-degree polynomial with integer coefficients and the coefficient to x^n is 1. Meanwhile, because $n \geq 4$, $(x+1)(x+2)\cdot(x+n)$ is a product of $n \geq 4$ consecutive number thus must be a multiple of 4. This means $f(x) \equiv 2 \pmod{4}$.

Practice 143

(Fermat's little theorem) Show that $a^p \equiv a \pmod{p}$ holds if p is a prime.

(ref: 4235)

First, $0^p \equiv 0 \pmod{p}$ and $1^p \equiv 1 \pmod{p}$ plainly hold.

By the conclusion of practice 97, we have

$$(x+y)^p \equiv x^p + y^p \pmod{p}$$

Setting $x = y = 1$ leads to

$$2^p \equiv 1^p + 1^p \equiv 2 \pmod{p}$$

Setting $x = 1$ and $y = 2$ leads to

$$3^p \equiv 1^p + 2^p \equiv 1 + 2 \equiv 3 \pmod{p}$$

Repeating this process will lead to the conclusion that

$$a^p \equiv a \pmod{p}$$

holds for all $a = 0, 12, 3, \cdots, p-1$. For any $a \geq p$, $a \pmod{p}$ msut equal to one of $\{0, 1, \cdots, p-1\}$. Hence we conclude $a^p \equiv a \pmod{p}$ holds for any a.

Practice 144

An integer in the form of $F_n = 2^{2^n} + 1$ where integer $n \geq 1$ is called a Fermat's number. Let d_n be any divisor of F_n. Show that $d_n \equiv 1 \pmod{2^{n+1}}$.

(ref: 4237)

Because any divisor of F_n is a product of some prime divisors of F_n, therefore it is sufficient to show that any prime divisor of F_n is congruent to 1 modulo 2^{n+1}.

Let p be any prime divisor of F_n. It is obvious that $p \neq 2$. Then, because $p \mid F_n = 2^{2^n} + 1$, we have

$$2^{2^n} \equiv -1 \pmod{p} \implies 2^{2^{n+1}} \equiv 1 \pmod{p}$$

Let r be the order of 2 modulo p. Then $r \mid 2^{n+1}$ which means r is some powers of 2. Let $r = 2^m$ where $0 \leq m \leq n+1$.

If $m \leq n$, then $2^{2^m} \equiv 1 \pmod{p}$. Continuously taking square of this relation will eventually yield $2^{2^n} \equiv 1 \pmod{p}$. Considering this together with the first equation above, i.e. $2^{2^n} \equiv -1 \pmod{p}$, will force $p = 2$. This contradicts to the fact that $p \neq 2$. Hence, we conclude m has to equal $(n+1)$, or $r = 2^m = 2^{n+1}$.

Meanwhile, by Fermat's little theorem, we have $2^{p-1} \equiv 1 \pmod{p}$. Therefore

$$r \mid (p-1) \implies 2^{n+1} \mid (p-1) \implies p \equiv 1 \pmod{2^{n+1}}$$

Practice 145

Assume positive integer $n > 1$ satisfies $n \mid (2^n + 1)$, prove n is a multiple of 3.

(ref: 4238)

Clearly, n is odd because $(2^n + 1)$ is odd. Let p be the minimal prime divisor of n. We will show that $p = 3$ which implies $3 \mid n$.

Let r be the order of 2 modulo p, then

$$2^r \equiv 1 \pmod{p}$$

Because $n \mid (2^n + 1)$ and $p \mid n$, we have

$$2^n \equiv -1 \pmod{p} \implies 2^{2n} = (2^n)^2 \equiv 1 \pmod{p}$$

Meanwhile, because $p \geq 3$ is odd, 2^{p-1} cannot be divisible by p. Applying Fermat's little theorem leads to

$$2^{p-1} \equiv 1 \pmod{p}$$

Because r is the order of 2 modulo p, the previous three relations imply $r \mid 2n$ and $r \mid (p-1)$. Therefore we conclude $r \mid (2n, p-1)$.

Now we show that $(2n, p-1) = 2$. If so, $r \mid 2 \implies r = 2$. In this case the first equation above will lead to the desired result of $p = 3$.

Because p is odd, therefore $(p-1)$ is even. This means that $2 \mid (2n, p-1)$. However, because n is odd, $4 \nmid 2n$ which implies $4 \nmid (2n, p-1)$.

Now, we claim that no odd prime can divide $(2n, p-1)$. This is because assuming there exists such an odd prime q, then $q \mid 2n$ implies $q \mid n$. Meanwhile $q \mid (p-1)$ implies $q < p$. This contradicts to the earlier assumption p is the smallest prime divisor of n.

Therefore, $(2n, p-1)$ is a multiple of 2, but not 4 or any odd prime. This means $(2n, p-1) = 2$.

Practice 146

Let p be an odd prime, and $n = \frac{2^{2p}-1}{3}$ in an integer. Prove $2^{n-1} \equiv 1 \pmod{n}$.

(ref: 4239)

The given condition yields

$$n - 1 = \frac{2^{2p} - 4}{3} \implies 3 \times (n-1) = 4 \times (2^{p-1}+1)(2^{p-1}-1)$$

Because p is an odd prime, Fermat's little theorem gives $2^{p-1} \equiv 1 \pmod{p}$, or $p \mid (2^{p-1}-1)$. Then, the right side of the above equation must be a multiple of $2p$. But because p is prime, $2p$ cannot be divisible by 3. Hence, it must hold that

$$2p \mid (n-1) \implies (2^{2p}-1) \mid (2^{n-1}-1)$$

Meanwhile, the given condition implies $n \mid (2^{2p}-1)$. Therefore it must be true that $n \mid (2^{n-1}-1)$ or $2^{n-1} \equiv 1 \pmod{n}$.

Practice 147

Show that there exists an infinite number of squares in the form of $(n \cdot 2^k - 7)$ where n and k are both positive integers.

(ref: 4251)

This is equivalent to showing this MOD equation has infinitely many solutions for any given positive integer k:

$$m^2 + 7 \equiv 0 \pmod{2^k}$$

When $k = 1, 2, 3$, any integer $m \equiv 1 \pmod{2^k}$ will satisfy the above MOD relation.

Assume that for $k \geq 3$, there exists m such that $m^2 + 7 \pmod 0$ $\pmod{2^k}$. Clearly, m is odd.

Then, for the case of $(k+1)$, let's consider the number $\left(m + a \cdot 2^{k-1}\right)$ where a is a positive integer:

$$\left(m + a \cdot 2^{k-1}\right)^2 + 7 \equiv m^2 + am \cdot 2^k + 7 \equiv 2^k(am + b) \pmod{2^{k+1}}$$

where $b = \frac{m^2+7}{2^k}$. By the assumption of the case k, we know b is an integer. Now, because m is odd, as long as a and b have the same parity, $(am + b)$ will be even. This means

$$2^k(am+b) \equiv p \pmod{2^{k+1}} \implies \left(m + a \cdot 2^{k-1}\right)^2 + 7 \equiv 0 \pmod{2^{k+1}}$$

Hence, by the principle of mathematical induction, we claim $m^2 + 7 \equiv 0 \pmod{2^k}$ is always solvable. This means there exist infinitely many square numbers in the form of $(n \cdot 2^k + 7)$.

Practice 148

Show that the number $(2n^{3k} + 4n^k + 10)$ cannot be a product of consecutive integers for any positive integers n and k.

(ref: 4256)

Let $M = 2n^{3k} + 4n^k + 10$. If it is a product of any three consecutive integers, then it must be a multiple of 3 because one of these three integers must be a multiple of 3. However, we claim that $M \equiv 1 \pmod 3$ which means it is not a multiple of 3. To show this, let $m = n^k$, then $M = 2m^3 + 4m + 10$.

- If $m \equiv 0 \pmod 3$, then $N \equiv 0 + 0 + 10 \equiv 1 \pmod 3$

- If $m \equiv 1 \pmod 3$, then $N \equiv 2+4+10 \equiv 1 \pmod 3$
- If $m \equiv -1 \pmod 3$, then $N \equiv -2-4+10 \equiv 1 \pmod 3$

Meanwhile, we can show M cannot be a product of two consecutive integers either. In order to show this, let's exam all the possible cases below where p is an integer.

- $3p(3p+1) \equiv 0 \pmod 3$
- $(3p+1)(3p+2) \equiv 2 \pmod 3$
- $(3p+2)(3p+3) \equiv 0 \pmod 3$

But we know $M \equiv 1 \pmod 3$. Therefore M cannot be a product of two consecutive integers.

Hence, we conclude the claim holds.

Practice 149

Let integers $l > m > n$ be the side lengths of a triangle satisfying $\left\{\frac{3^l}{10^4}\right\} = \left\{\frac{3^m}{10^4}\right\} = \left\{\frac{3^n}{10^4}\right\}$ where function $\{x\}$ returns the decimal part of real number x. Find the least possible value of this triangle's perimeter.

(ref: 4259)

The given condition implies that the last four digits of 3^l, 3^m and 3^n are the same.

$$\therefore \quad 10^4 \mid (3^m - 3^n) = 3^n\left(3^{m-n} - 1\right)$$
$$\implies 3^{m-n} \equiv 1 \pmod{10^4}$$
$$\implies m - n \equiv 0 \pmod{500}$$

The last conclusion holds because the order of 3 modulo 10^4 is 500. Similarly, we must have $l - m \equiv 0 \pmod{500}$ as well.

Let $m - n = 500a$ and $l - m = 500b$ where a and b are positive integers. By triangular inequality, we have

$$n > l - m = 500b \implies n_{min} = 500 \times 1 + 1 = 501$$

Accordingly, $m_{min} = 1001$ and $l_{min} = 1501$. Therefore, the minimal perimeter equals $501 + 1001 + 1501 = \boxed{3003}$.

Practice 150

Solve $x^{22} + x^{11} \equiv 2 \pmod{11}$.

(ref: 4168)

Because 2 is co-prime to 11, we find that any solution x must be co-prime to 11.

By Fermat's little theorem, we have $x^{10} \equiv 1 \pmod{11}$. It follows that

$$x^{22} + x^{11} \equiv x^2 + x \equiv x^2 + 12x + 2 \equiv (x+6)^2 - 36 \equiv 2 \pmod{11}$$

This is equivalent to $(x+6)^2 \equiv 5 \pmod{11}$. Testing $(x+6) = 0$, ± 1, ± 2, ± 3, ± 4, and ± 5 finds $(x+6) = \pm 4$ are solutions. Hence, the solutions to the given equation are $x \equiv 1 \pmod{11}$ or $x \equiv 9 \pmod{11}$.

Practice 151

Let sequence $g(n)$ satisfy $g(1) = 0, g(2) = 1, g(n+2) = g(n+1) + g(n) + 1$ where $n \geq 1$. Show that if n is a prime greater than 5, then $n \mid g(n)[g(n) + 1]$.

(ref: 2699 - IMO)

Let $f(n) = g(n) + 1$, then $f(1) = 1$, $f(2) = 2$, and $f(n+2) = f(n+1) + f(n)$. The solution to this sequence is

$$f(n) = \frac{1}{\sqrt{5}} \cdot \left[\left(\frac{1+\sqrt{5}}{2}\right)^{n+1} - \left(\frac{1-\sqrt{5}}{2}\right)^{n+1} \right]$$

$$= \frac{1}{2^n} \cdot \left(C_{n+1}^1 + 5C_{n+1}^3 + 5^2 C_{n+1}^5 + \cdots + 5^{\frac{n-1}{2}} C_{n+1}^n \right)$$

Because n is a prime greater than 5, therefore $(2, n) = 1 \implies (2^n, n) = 1$ and $n \mid C_{n+1}^i$ where $3 \le i \le n-1$. Therefore, the previous relation will lead to

$$2^n f(n) \equiv C_{n+1}^1 + 5^{\frac{n-1}{2}} C_{n+1}^n \equiv (n+1)\left(1 + 5^{\frac{n-1}{2}}\right) \pmod{n}$$

It follows that

$$2^n [f(n) - 1] \equiv 1 + 5^{\frac{n-1}{2}} - 2^n \equiv -1 + 5^{\frac{n-1}{2}} \pmod{n}$$

by Fermat Little Theorem. Multiplying the last two equations gives

$$2^{2n} f(n)[f(n) - 1] \equiv 5^{n-1} - 1 \equiv 0 \pmod{n}$$

by Fermat Little Theorem again. Therefore

$$f(n)[f(n) - 1] \equiv 0 \pmod{5} \implies g(n)[g(n) + 1] \equiv 0 \pmod{5}$$

Practice 152

Let p be an odd prime. Show that

$$\sum_{j=0}^{p} C_p^j C_{p+j}^j \equiv 2^p + 1 \pmod{p^2}$$

(ref: 4172 - Putnam)

We claim the left side is the coefficient of x^p in the expanded form of $(1+x)^p (2+x)^p$.

This is because $C_{p+j}^j = C_{p+j}^p$, therefore the left side is the coefficient of x^p in the following polynomial

$$\sum_{j=0}^{p} \left(C_p^j \sum_{k=0}^{p+j} C_{p+j}^k \cdot x^k \right) = \sum_{j=0}^{p} C_p^j (1+x)^{p+j}$$
$$= (1+x)^p \sum_{j=0}^{p} C_p^j (1+x)^j = (1+x)^p (2+x)^p$$

On the other hand, expanding directly gives us the coefficient of x^p be

$$\sum_{k=0}^{p} C_p^k C_p^{p-k} \cdot 2^k = \sum_{k=0}^{p} \left(\frac{p!}{k!(p-k)!} \right)^2 \cdot 2^k$$

By practice 83, we find $p \mid C_n^k$ when $0 < k < p$ which means $p^2 \mid \left(C_n^k \right)^2 = \left(\frac{p!}{k!(p-k)!} \right)^2$. Hence, all the terms in the above expressions are multiples of p^2, except when $k = 0$ and $k = p$. It follows that its modulo p^2 equals $(1 + 2^p)$.

Practice 153

Show that if the equation $a^2 + 1 \equiv 0 \pmod{p}$ is solvable for some a, then p can be represented as a sum of two squares.

(ref: 4187)

If a is a solution to this equation, then we must have $p \nmid a$. This is because if $p \mid a$, then $a^2 + 1 \equiv 1 \pmod{p}$

By Thue's Theorem (practice 133), there exist positive integers x and y not exceeding $\lfloor \sqrt{p} \rfloor$ such that

$$ax \equiv \pm y \pmod{p} \implies a^2 x^2 \equiv y^2 \pmod{p}$$

Now multiplying $a^2 + 1 \equiv 0 \pmod{p}$ by x^2 gives

$$0 \equiv a^2 x^2 + x^2 \equiv x^2 + y^2 \pmod{p}$$

This means $x^2 + y^2 = kp$ where k is a positive integer. If we can show $k = 1$, then p is a sum of two squares. This is indeed the case because

$$x, y \leq \lfloor \sqrt{p} \rfloor \implies x^2 + y^2 < 2p \implies k < 2$$

Practice 154

Show that a prime $p > 2$ is a sum of two squares if and only if $p \equiv 1 \pmod 4$.

(ref: 4188)

If p is a sum of two squares, let $p = x^2 + y^2$. Because p must be odd, therefore one of x and y must be even and the other must be odd. Without loss of generality, let x be even and y be odd. Then it is easy to show that $x^2 \equiv 0 \pmod 4$ and $y^2 \equiv 1 \pmod 4$. Hence, $p = x^2 + y^2 \equiv 0 + 1 \equiv 1 \pmod 4$.

On the other hand, if $p \equiv 1 \pmod 4$. We are going to show that $a^2 + 1 = p$ is solvable for some integer a. If so, by practice 153, we know p is a sum of two squares. For this, let's consider factors of $(p-1)!$:

$$(p-1)! = 1 \cdot 2 \cdot 3 \cdots \left(\frac{p-1}{2}\right) \cdot \left(\frac{p+1}{2}\right) \cdots (p-2) \cdot (p-1)$$

Each factor in the first half can be paired with a corresponding one in the second half with negative congruent relationship:

$$\begin{aligned} 1 &\equiv -(p-1) &\pmod p \\ 2 &\equiv -(p-2) &\pmod p \\ &\cdots \\ \tfrac{p-1}{2} &\equiv -\tfrac{p+1}{2} &\pmod p \end{aligned}$$

Therefore, we have

$$(p-1)! \equiv (-1)^{\frac{p-1}{2}} \left(1 \cdot 2 \cdot 3 \cdots \left(\frac{p-1}{2}\right)\right)^2 \pmod p$$

Meanwhile, by Wilson's theorem, we have $(p-1)! \equiv -1 \pmod p$. Thus

$$-1 \equiv (-1)^{\frac{p-1}{2}} \left(1 \cdot 2 \cdot 3 \cdots \left(\frac{p-1}{2}\right)\right)^2 \pmod p$$

or
$$(-1)^{\frac{p+1}{2}} \equiv \left(1 \cdot 2 \cdot 3 \cdots \left(\frac{p-1}{2}\right)\right)^2 \pmod{p}$$

As $p \equiv 1 \pmod{4}$, let $p = 4k + 1$ where k is an integer. Then $\frac{p+1}{2} = 2k + 1$ is odd. Then the previous relation can be written as

$$\left(1 \cdot 2 \cdot 3 \cdots \left(\frac{p-1}{2}\right)\right)^2 + 1 \equiv 0 \pmod{p}$$

Let $a = 1 \cdot 2 \cdot 3 \cdots \left(\frac{p-1}{2}\right)$, we have $a^2 + 1 \equiv 0 \pmod{p}$ which is what we are looking for.

Practice 155

(Two Squares Theorem) Show that a positive integer n is a sum of two squares if and only if each prime factor p of n such that $p \equiv 3 \pmod{4}$ occurs to an even power in the prime factorization of n.

(ref: 4190)

First let $n = s^2 m$ where both s and m are positive integers, and none of m's divisor greater than 1 is a square.

If every prime factor $p \equiv 3 \pmod{4}$ occurs an even power in the prime factorization of n, all prime factors $p > 2$ of m must be in the form of $p \equiv 1 \pmod{4}$. By practice 154, we know all of these prime factors can be written as a sum of two squares. Meanwhile, it is clear that $2 = 1^2 + 1^2$. Therefore m can be written as a product of several sum of squares.

We know that a product of two sums of squares can be written as a sum of two squares (see practice 95). Thus, m can always be expressed as a sum of two squares. Let $m = x^2 + y^2$ where both x and y are integers. This implies $n = s^2 m = (sx)^2 + (sy)^2$.

On the other hand, if n is a sum of two squares, let it be $x^2 + y^2$. If $m = 1$, then $n = s^2$. In this case, all prime factors occurs even power in the prime factorization of n. Therefore it is sufficient to show that if $m > 1$, every odd prime factor p of m must satisfy $p \equiv 1 \pmod 4$.

Let d be the greatest common divisor of x and y. Then there exist two co-prime integers x_1 and y_1 such that $x = dx_1$ and $y = dy_1$. It follows that

$$n = d^2(x_1^2 + y_1^2) \implies \frac{s^2 m}{d^2} = x_1^2 + y_1^2$$

Because m does not have any square divisor, we must have $d^2 \mid s^2$. Let's assume $s^2 = t \cdot d^2$ where t is an integer. Then

$$tm = x_1^2 + y_1^2$$

If p is a prime factor of m, then we must have $p \mid x_1^2 + y_1^2$ or

$$x_1^2 + y_1^2 \equiv 0 \pmod p \implies x_1^2 \equiv -y_1^2 \pmod p$$

If $p \not\equiv 1 \pmod 4$, then $p \equiv 3 \pmod 4$ or $p = 4k+3$ where k is an integer. Noting that $p - 1 = 2 \times (2k+1)$, we have

$$\begin{aligned} x_1^2 &\equiv -y_1^2 & \pmod p \\ (x_1^2)^{2k+1} &\equiv -(y_1^2)^{2k+1} & \pmod p \\ x_1^{p-1} &\equiv -y_1^{p-1} & \pmod p \end{aligned}$$

We note that p cannot divide either x_1 or y_1. If this conclusion does not hold, then p must divide both x_1 and y_1 because $p \mid x_1^2 + y_1^2$. This will contradict to the fact that x_1 and y_1 are co-prime. If p divides neither x_1 nor y_1, then they must be co-prime because p is a prime. In this case, by Fermat little theorem, we have

$$x_1^{p-1} \equiv y_1^{p-1} \equiv 1 \pmod p$$

Setting this to the early result of $x_1^{p-1} \equiv y_1^{p-1} \equiv 1 \pmod p$ will yield $1 \equiv -1 \pmod p$. This will force $p = 2$ which contradicts to the assumption that p is an odd prime. This means $p \equiv 1 \pmod 4$ has to be held.